SSAT Upper-Level

Subject Test Mathematics

Student Practice Workbook

+ Two Full-Length SSAT Upper-Level Math Tests

SCAN ME

Math Notion

www.MathNotion.com

SSAT Upper-Level Subject Test Mathematics

Published in the United State of America By

The Math Notion

Web: WWW.MathNotion.com

Email: info@Mathnotion.com

ISBN: 978-1-63620-062-0

The Math Notion

Michael Smith has been a math instructor for over a decade now. He launched the Math Notion. Since 2006, we have devoted our time to both teaching and developing exceptional math learning materials. As a test prep company, we have worked with thousands of students. We have used the feedback of our students to develop a unique study program that can be used by students to drastically improve their math scores fast and effectively. We have more than a thousand Math learning books including:

– **SAT Math Prep**

– **ACT Math Prep**

– **SSAT/ISEE Math Prep**

– **Accuplacer Math Prep**

– **Common Core Math Prep**

–**many Math Education Workbooks, Study Guides, Practice and Exercise Books**

As an experienced Math test preparation company, we have helped many students raise their standardized test scores—and attend the colleges of their dreams: We tutor online and in person, we teach students in large groups, and we provide training materials and textbooks through our website and through Amazon.

You can contact us via email at:

info@Mathnotion.com

Get the Targeted Practice You Need to Ace the SSAT Upper-Level Math Test!

SSAT Upper-Level Subject Test Mathematics includes easy-to-follow instructions, helpful examples, and plenty of math practice problems to assist students to master each concept, brush up their problem-solving skills, and create confidence.

The SSAT Upper-Level math practice book provides numerous opportunities to evaluate basic skills along with abundant remediation and intervention activities. It is a skill that permits you to quickly master intricate information and produce better leads in less time.

Students can boost their test-taking skills by taking the book's two practice SSAT Upper-Level Math exams. All test questions answered and explained in detail.

Important Features of the SSAT Upper-Level Math Book:

- A **complete review** of SSAT Upper-Level math test topics,
- Over 2,500 practice problems covering all topics tested,
- The most important concepts you need to know,
- Clear and concise, easy-to-follow sections,
- Well designed for enhanced learning and interest,
- Hands-on experience with all question types,
- **2 full-length practice tests** with detailed answer explanations,
- Cost-Effective Pricing,

Powerful math exercises to help you avoid traps and pacing yourself to beat the SSAT Upper-Level test. Students will gain valuable experience and raise their confidence by taking math practice tests, learning about test structure, and gaining a deeper understanding of what is tested on the SSAT Upper-Level Math. If ever there was a book to respond to the pressure to increase students' test scores, this is it.

WWW.MathNotion.COM

... So Much More Online!

✓ FREE Math Lessons

✓ More Math Learning Books!

✓ Mathematics Worksheets

✓ Online Math Tutors

For a PDF Version of This Book

SCAN ME

Please Visit WWW.MathNotion.com

Contents

Chapter 1 :

Integers and Number Theory

Topics that you will practice in this chapter:

- ✓ Rounding
- ✓ Whole Number Addition and Subtraction
- ✓ Whole Number Multiplication and Division
- ✓ Rounding and Estimates
- ✓ Adding and Subtracting Integers
- ✓ Multiplying and Dividing Integers
- ✓ Order of Operations
- ✓ Ordering Integers and Numbers
- ✓ Integers and Absolute Value
- ✓ Factoring Numbers
- ✓ Greatest Common Factor (GCF)
- ✓ Least Common Multiple (LCM)

"Wherever there is number, there is beauty." –Proclus

Rounding

✎ **Round each number to the nearest ten.**

1) 42 = ____

2) 88 = ____

3) 24 = ____

4) 57 = ____

5) 19 = ____

6) 25 = ____

7) 93 = ____

8) 71 = ____

9) 48 = ____

10) 81 = ____

11) 58 = ____

12) 87 = ____

✎ **Round each number to the nearest hundred.**

13) 198 = ____

14) 387 = ____

15) 816 = ____

16) 101 = ____

17) 321 = ____

18) 433 = ____

19) 579 = ____

20) 825 = ____

21) 580 = ____

22) 868 = ____

23) 480 = ____

24) 287 = ____

✎ **Round each number to the nearest thousand.**

25) 1,382 = ____

26) 3,420 = ____

27) 4,254 = ____

28) 6,861 = ____

29) 9,099 = ____

30) 22,980 = ____

31) 45,188 = ____

32) 16,808 = ____

33) 52,866 = ____

34) 85,190 = ____

35) 70,990 = ____

36) 26,869 = ____

Rounding and Estimates

✎ **Estimate the sum by rounding each number to the nearest ten.**

1) $13 + 22 =$ _____

2) $71 + 23 =$ _____

3) $61 + 58 =$ _____

4) $56 + 85 =$ _____

5) $368 + 249 =$ _____

6) $330 + 903 =$ _____

7) $471 + 293 =$ _____

8) $1,950 + 2,655 =$ _____

✎ **Estimate the product by rounding each number to the nearest ten.**

9) $32 \times 71 =$ _____

10) $12 \times 33 =$ _____

11) $31 \times 83 =$ _____

12) $19 \times 11 =$ _____

13) $42 \times 76 =$ _____

14) $63 \times 34 =$ _____

15) $19 \times 31 =$ _____

16) $59 \times 71 =$ _____

✎ **Estimate the sum or product by rounding each number to the nearest ten.**

17)
$$\begin{array}{r} 29 \\ \times\ 12 \\ \hline \end{array}$$

18)
$$\begin{array}{r} 37 \\ \times\ 26 \\ \hline \end{array}$$

19)
$$\begin{array}{r} 48 \\ +\ 82 \\ \hline \end{array}$$

20)
$$\begin{array}{r} 65 \\ +44 \\ \hline \end{array}$$

21)
$$\begin{array}{r} 37 \\ \times\ 14 \\ \hline \end{array}$$

22)
$$\begin{array}{r} 71 \\ +\ 32 \\ \hline \end{array}$$

Adding and Subtracting Integers

✎ **Find each sum.**

1) $14 + (-6) =$

2) $(-13) + (-20) =$

3) $5 + (-28) =$

4) $50 + (-12) =$

5) $(-7) + (-15) + 3 =$

6) $30 + (-14) + 8 =$

7) $40 + (-10) + (-14) + 17 =$

8) $(-15) + (-20) + 13 + 35 =$

9) $40 + (-20) + (38 - 29) =$

10) $28 + (-12) + (30 - 12) =$

✎ **Find each difference.**

11) $(-18) - (-7) =$

12) $25 - (-14) =$

13) $(-20) - 36 =$

14) $34 - (-19) =$

15) $51 - (30 - 21) =$

16) $17 - (5) - (-24) =$

17) $(35 + 20) - (-46) =$

18) $48 - 16 - (-8) =$

19) $62 - (28 + 17) - (-15) =$

20) $58 - (-23) - (-31) =$

21) $19 - (-8) - (-13) =$

22) $(19 - 24) - (-14) =$

23) $27 - 33 - (-21) =$

24) $58 - (32 + 24) - (-9) =$

25) $36 - (-30) + (-17) =$

26) $27 - (-42) + (-31) =$

Multiplying and Dividing Integers

✎ **Find each product.**

1) $(-9) \times (-5) =$

2) $(-3) \times 9 =$

3) $8 \times (-12) =$

4) $(-7) \times (-20) =$

5) $(-3) \times (-5) \times 6 =$

6) $(14 - 3) \times (-8) =$

7) $12 \times (-9) \times (-3) =$

8) $(140 + 10) \times (-2) =$

9) $10 \times (-12 + 8) \times 3 =$

10) $(-8) \times (-5) \times (-10) =$

✎ **Find each quotient.**

11) $42 \div (-7) =$

12) $(-48) \div (-6) =$

13) $(-40) \div (-8) =$

14) $54 \div (-2) =$

15) $152 \div 19 =$

16) $(-144) \div (-12) =$

17) $180 \div (-10) =$

18) $(-312) \div (-12) =$

19) $221 \div (-13) =$

20) $(-126) \div (6) =$

21) $(-161) \div (-7) =$

22) $-266 \div (-14) =$

23) $(-120) \div (-4) =$

24) $270 \div (-18) =$

25) $(-208) \div (-8) =$

26) $(135) \div (-15) =$

Order of Operations

✑ **Evaluate each expression.**

1) $7 + (5 \times 4) =$

2) $14 - (3 \times 6) =$

3) $(19 \times 4) + 16 =$

4) $(16 - 7) - (8 \times 2) =$

5) $27 + (18 \div 3) =$

6) $(18 \times 8) \div 6 =$

7) $(32 \div 4) \times (-2) =$

8) $(9 \times 4) + (32 - 18) =$

9) $24 + (4 \times 3) + 7 =$

10) $(36 \times 3) \div (2 + 2) =$

11) $(-7) + (12 \times 3) + 11 =$

12) $(8 \times 5) - (24 \div 6) =$

13) $(7 \times 6 \div 3) - (12 + 9) =$

14) $(13 + 5 - 14) \times 3 - 2 =$

15) $(20 - 14 + 30) \times (64 \div 4) =$

16) $32 + \big(28 - (36 \div 9)\big) =$

17) $(7 + 6 - 4 - 7) + (15 \div 5) =$

18) $(85 - 20) + (20 - 18 + 7) =$

19) $(20 \times 2) + (14 \times 3) - 22 =$

20) $18 + 5 - (30 \times 3) + 20 =$

Ordering Integers and Numbers

✍ **Order each set of integers from least to greatest.**

1) $8, -10, -5, -3, 4$ ___, ___, ___, ___, ___, ___

2) $-10, -18, 6, 14, 27$ ___, ___, ___, ___, ___, ___

3) $15, -8, -21, 21, -23$ ___, ___, ___, ___, ___, ___

4) $-14, -40, 23, -12, 47$ ___, ___, ___, ___, ___, ___

5) $59, -54, 32, -57, 36$ ___, ___, ___, ___, ___, ___

6) $68, 26, -19, 47, -34$ ___, ___, ___, ___, ___, ___

✍ **Order each set of integers from greatest to least.**

7) $18, 36, -16, -18, -10$ ___, ___, ___, ___, ___, ___

8) $27, 34, -12, -24, 94$ ___, ___, ___, ___, ___, ___

9) $50, -21, -13, 42, -2$ ___, ___, ___, ___, ___, ___

10) $37, 46, -20, -16, 86$ ___, ___, ___, ___, ___, ___

11) $-18, 88, -26, -59, 75$ ___, ___, ___, ___, ___, ___

12) $-65, -30, -25, 3, 14$ ___, ___, ___, ___, ___, ___

Integers and Absolute Value

✏️ **Write absolute value of each number.**

1) $|-2| =$

2) $|-27| =$

3) $|-20| =$

4) $|14| =$

5) $|6| =$

6) $|-55| =$

7) $|16| =$

8) $|2| =$

9) $|54| =$

10) $|-4| =$

11) $|-11|$

12) $|88| =$

13) $|0| =$

14) $|79| =$

15) $|-32| =$

16) $|-17| =$

17) $|42| =$

18) $|-46| =$

19) $|1| =$

20) $|-40| =$

✏️ **Evaluate the value.**

21) $|-5| - \dfrac{|-21|}{7} =$

22) $14 - |3 - 15| - |-4| =$

23) $\dfrac{|-32|}{4} \times |-4| =$

24) $\dfrac{|7 \times (-3)|}{7} \times \dfrac{|-19|}{3} =$

25) $|4 \times (-5)| + \dfrac{|-40|}{5} =$

26) $\dfrac{|-45|}{9} \times \dfrac{|-24|}{12} =$

27) $|-12 + 8| \times \dfrac{|-7 \times 7|}{7} =$

28) $\dfrac{|-11 \times 2|}{4} \times |-16| =$

Factoring Numbers

✎ **List all positive factors of each number.**

1) 9

2) 16

3) 24

4) 30

5) 26

6) 46

7) 20

8) 68

9) 28

10) 98

11) 14

12) 54

13) 55

14) 18

15) 63

16) 34

17) 50

18) 62

19) 95

20) 64

21) 70

22) 45

23) 22

24) 65

Greatest Common Factor

✍ **Find the GCF for each number pair.**

1) 6, 2

2) 4, 5

3) 3, 12

4) 7, 3

5) 5, 10

6) 8, 48

7) 6, 18

8) 9, 15

9) 12, 18

10) 4, 36

11) 6, 10

12) 28, 52

13) 25, 10

14) 22, 24

15) 9, 54

16) 8, 54

17) 42, 14

18) 16, 40

19) 9, 2, 3

20) 5, 15, 10

21) 7, 9, 2

22) 16, 64

23) 30, 48

24) 36, 63

Least Common Multiple

✑ **Find the LCM for each number pair.**

1) 6, 9

2) 15, 45

3) 16, 40

4) 12, 36

5) 18, 27

6) 14, 42

7) 6, 30

8) 8, 56

9) 7, 21

10) 8, 20

11) 15, 25

12) 7, 9

13) 4, 11

14) 8, 28

15) 28, 56

16) 40, 50

17) 12, 13

18) 22, 11

19) 36, 20

20) 15, 35

21) 18, 81

22) 30, 54

23) 18, 45

24) 75, 25

Answers of Worksheets

Rounding

1) 40	10) 80	19) 600	28) 7,000
2) 90	11) 60	20) 800	29) 9,000
3) 20	12) 90	21) 600	30) 23,000
4) 60	13) 200	22) 900	31) 45,000
5) 20	14) 400	23) 500	32) 17,000
6) 30	15) 800	24) 300	33) 53,000
7) 90	16) 100	25) 1,000	34) 85,000
8) 70	17) 300	26) 3,000	35) 71,000
9) 50	18) 400	27) 4,000	36) 27,000

Rounding and Estimates

1) 30	7) 760	13) 3,200	19) 130
2) 90	8) 4,610	14) 1,800	20) 110
3) 120	9) 2,100	15) 600	21) 400
4) 150	10) 300	16) 4,200	22) 100
5) 620	11) 2,400	17) 300	
6) 1,230	12) 200	18) 1,200	

Adding and Subtracting Integers

1) 8	8) 13	15) 42	22) 9
2) −33	9) 29	16) 36	23) 15
3) −23	10) 34	17) 101	24) 11
4) 38	11) −11	18) 40	25) 49
5) −19	12) 39	19) 32	26) 38
6) 24	13) −56	20) 112	
7) 33	14) 53	21) 40	

Multiplying and Dividing Integers

1) 45	6) −88	11) −6	16) 12
2) −27	7) 324	12) 8	17) −18
3) −96	8) −300	13) 5	18) 26
4) 140	9) −120	14) −27	19) −17
5) 90	10) −400	15) 8	20) −21

21) 23	23) 30	25) 26
22) 19	24) −15	26) −9

Order of Operations

1) 27	6) 24	11) 40	16) 56
2) −4	7) −16	12) 36	17) 5
3) 92	8) 50	13) −7	18) 74
4) −7	9) 43	14) 10	19) 60
5) 33	10) 27	15) 576	20) −47

Ordering Integers and Numbers

1) −10, −5, −3, 4, 8	7) 36, 18, −10, −16, −18
2) −18, −10, 6, 14, 27	8) 94, 34, 27, −12, −24
3) −23, −21, −8, 15, 21	9) 50, 42, −2, −13, −21
4) −40, −14, −12, 23, 47	10) 86, 46, 37, −16, −20
5) −57, −54, 32, 36, 59	11) 88, 75, −18, −26, −59
6) −34, −19, 26, 47, 68	12) 14, 3, −25, −30, −65

Integers and Absolute Value

1) 2	8) 2	15) 32	22) −2
2) 27	9) 54	16) 17	23) 32
3) 20	10) 4	17) 42	24) 19
4) 14	11) 11	18) 46	25) 28
5) 6	12) 88	19) 1	26) 10
6) 55	13) 0	20) 40	27) 28
7) 16	14) 79	21) 2	28) 88

Factoring Numbers

1) 1, 3, 9	9) 1, 2, 4, 7, 14, 28	17) 1, 2, 5, 10, 25, 50
2) 1, 2, 4, 8, 16	10) 1, 2, 7, 14, 49, 98	18) 1, 2, 31, 62
3) 1, 2, 3, 4, 6, 8, 12, 24	11) 1, 2, 7, 14	19) 1, 5, 19, 95
4) 1, 2, 3, 5, 6, 10, 15, 30	12) 1, 2, 3, 6, 9, 18, 27, 54	20) 1, 2, 4, 8, 16, 32, 64
5) 1, 2, 13, 26	13) 1, 5, 11, 55	21) 1, 2, 5, 7, 10, 14, 35, 70
6) 1, 2, 23, 46	14) 1, 2, 3, 6, 9, 18	22) 1, 3, 5, 9, 15, 45
7) 1, 2, 4, 5, 10, 20	15) 1, 3, 7, 9, 21, 63	23) 1, 2, 11, 22
8) 1, 2, 4, 17, 34, 68	16) 1, 2, 17, 34	24) 1, 5, 13, 65

Greatest Common Factor

1) 2	7) 6	13) 5	19) 1
2) 1	8) 3	14) 2	20) 5
3) 3	9) 6	15) 9	21) 1
4) 1	10) 4	16) 2	22) 16
5) 5	11) 2	17) 14	23) 6
6) 8	12) 4	18) 8	24) 9

Least Common Multiple

1) 18	7) 30	13) 44	19) 180
2) 45	8) 56	14) 56	20) 105
3) 80	9) 21	15) 56	21) 162
4) 36	10) 40	16) 200	22) 270
5) 54	11) 75	17) 156	23) 90
6) 42	12) 63	18) 22	24) 75

Chapter 2 :

Fractions and Decimals

Topics that you will practice in this chapter:

- ✓ Simplifying Fractions
- ✓ Adding and Subtracting Fractions
- ✓ Multiplying and Dividing Fractions
- ✓ Adding and Subtract Mixed Numbers
- ✓ Multiplying and Dividing Mixed Numbers
- ✓ Adding and Subtracting Decimals
- ✓ Multiplying and Dividing Decimals
- ✓ Comparing Decimals
- ✓ Rounding Decimals

"A Man is like a fraction whose numerator is what he is and whose denominator is what he thinks of himself. The larger the denominator, the smaller the fraction." –Tolstoy

Simplifying Fractions

✎ **Simplify each fraction to its lowest terms.**

1) $\frac{5}{10} =$

2) $\frac{28}{35} =$

3) $\frac{27}{36} =$

4) $\frac{40}{80} =$

5) $\frac{14}{56} =$

6) $\frac{32}{48} =$

7) $\frac{52}{65} =$

8) $\frac{15}{60} =$

9) $\frac{80}{160} =$

10) $\frac{55}{77} =$

11) $\frac{28}{112} =$

12) $\frac{32}{64} =$

13) $\frac{63}{72} =$

14) $\frac{81}{90} =$

15) $\frac{35}{105} =$

16) $\frac{25}{70} =$

17) $\frac{80}{280} =$

18) $\frac{12}{81} =$

19) $\frac{36}{186} =$

20) $\frac{240}{540} =$

21) $\frac{70}{560} =$

✎ **Find the answer for each problem.**

22) Which of the following fractions equal to $\frac{3}{4}$? ____

    A. $\frac{60}{90}$
    B. $\frac{43}{104}$
    C. $\frac{48}{64}$
    D. $\frac{150}{300}$

23) Which of the following fractions equal to $\frac{5}{8}$? ____

    A. $\frac{125}{200}$
    B. $\frac{115}{200}$
    C. $\frac{50}{100}$
    D. $\frac{30}{90}$

24) Which of the following fractions equal to $\frac{3}{7}$? ____

    A. $\frac{58}{116}$
    B. $\frac{54}{126}$
    C. $\frac{270}{167}$
    D. $\frac{42}{63}$

Adding and Subtracting Fractions

✎ **Find the sum.**

1) $\frac{5}{9} + \frac{4}{9} =$

5) $\frac{1}{4} + \frac{3}{5} =$

9) $\frac{5}{7} + \frac{2}{3} =$

2) $\frac{1}{2} + \frac{1}{7} =$

6) $\frac{7}{8} + \frac{3}{8} =$

10) $\frac{7}{12} + \frac{3}{4} =$

3) $\frac{3}{8} + \frac{1}{4} =$

7) $\frac{1}{2} + \frac{7}{10} =$

11) $\frac{5}{6} + \frac{2}{5} =$

4) $\frac{3}{5} + \frac{1}{2} =$

8) $\frac{2}{5} + \frac{2}{3} =$

12) $\frac{1}{12} + \frac{2}{3} =$

✎ **Find the difference.**

13) $\frac{1}{3} - \frac{1}{6} =$

19) $\frac{5}{6} - \frac{1}{9} =$

25) $\frac{6}{7} - \frac{3}{4} =$

14) $\frac{3}{4} - \frac{1}{8} =$

20) $\frac{3}{4} - \frac{1}{6} =$

26) $\frac{4}{5} - \frac{1}{8} =$

15) $\frac{1}{2} - \frac{1}{3} =$

21) $\frac{7}{8} - \frac{1}{12} =$

27) $\frac{4}{7} - \frac{2}{35} =$

16) $\frac{1}{4} - \frac{1}{5} =$

22) $\frac{8}{15} - \frac{3}{5} =$

28) $\frac{9}{16} - \frac{2}{8} =$

17) $\frac{5}{8} - \frac{2}{3} =$

23) $\frac{3}{12} - \frac{1}{14} =$

29) $\frac{8}{9} - \frac{7}{18} =$

18) $\frac{1}{4} - \frac{1}{7} =$

24) $\frac{10}{13} - \frac{7}{26} =$

30) $\frac{1}{2} - \frac{4}{9} =$

Multiplying and Dividing Fractions

✏ **Find the value of each expression in lowest terms.**

1) $\frac{1}{5} \times \frac{15}{5} =$

2) $\frac{9}{12} \times \frac{4}{9} =$

3) $\frac{1}{16} \times \frac{8}{10} =$

4) $\frac{1}{24} \times \frac{8}{10} =$

5) $\frac{1}{5} \times \frac{1}{4} =$

6) $\frac{7}{9} \times \frac{1}{7} =$

7) $\frac{6}{7} \times \frac{1}{3} =$

8) $\frac{2}{8} \times \frac{2}{8} =$

9) $\frac{5}{8} \times \frac{3}{5} =$

10) $\frac{4}{7} \times \frac{1}{8} =$

11) $\frac{7}{15} \times \frac{5}{7} =$

12) $\frac{3}{10} \times \frac{5}{9} =$

✏ **Find the value of each expression in lowest terms.**

13) $\frac{1}{4} \div \frac{1}{8} =$

14) $\frac{1}{10} \div \frac{1}{5} =$

15) $\frac{3}{4} \div \frac{1}{5} =$

16) $\frac{1}{3} \div \frac{5}{6} =$

17) $\frac{1}{7} \div \frac{8}{42} =$

18) $\frac{3}{4} \div \frac{1}{6} =$

19) $\frac{2}{7} \div \frac{7}{13} =$

20) $\frac{1}{24} \div \frac{3}{16} =$

21) $\frac{7}{12} \div \frac{5}{6} =$

22) $\frac{22}{18} \div \frac{11}{9} =$

23) $\frac{9}{35} \div \frac{3}{7} =$

24) $\frac{2}{7} \div \frac{8}{21} =$

25) $\frac{1}{9} \div \frac{2}{5} =$

26) $\frac{5}{12} \div \frac{3}{5} =$

27) $\frac{3}{20} \div \frac{1}{6} =$

28) $\frac{8}{20} \div \frac{3}{4} =$

29) $\frac{5}{6} \div \frac{2}{9} =$

30) $\frac{5}{11} \div \frac{3}{4} =$

Adding and Subtracting Mixed Numbers

✍ **Find the sum.**

1) $3\frac{1}{3} + 2\frac{1}{6} =$

6) $5\frac{4}{15} + 3\frac{3}{5} =$

2) $4\frac{1}{2} + 3\frac{1}{2} =$

7) $2\frac{1}{3} + 4\frac{3}{7} =$

3) $3\frac{3}{8} + 1\frac{1}{8} =$

8) $3\frac{1}{2} + 4\frac{2}{5} =$

4) $2\frac{1}{4} + 2\frac{1}{3} =$

9) $5\frac{2}{5} + 6\frac{3}{7} =$

5) $3\frac{5}{6} + 2\frac{7}{12} =$

10) $8\frac{5}{16} + 6\frac{1}{12} =$

✍ **Find the difference.**

11) $3\frac{1}{4} - 1\frac{3}{4} =$

19) $14\frac{5}{6} - 11\frac{3}{5} =$

12) $6\frac{3}{5} - 4\frac{2}{5} =$

20) $18\frac{2}{7} - 14\frac{1}{5} =$

13) $4\frac{1}{3} - 3\frac{1}{9} =$

21) $9\frac{1}{3} - 4\frac{1}{4} =$

14) $7\frac{1}{7} - 5\frac{1}{2} =$

22) $6\frac{1}{8} - 4\frac{1}{16} =$

15) $5\frac{1}{3} - 2\frac{1}{12} =$

23) $19\frac{3}{8} - 15\frac{1}{3} =$

16) $8\frac{1}{5} - 4\frac{1}{3} =$

24) $11\frac{1}{9} - 8\frac{1}{8} =$

17) $9\frac{1}{4} - 6\frac{1}{8} =$

25) $17\frac{1}{7} - 11\frac{1}{5} =$

18) $11\frac{7}{15} - 8\frac{3}{5} =$

26) $16\frac{2}{9} - 9\frac{5}{7} =$

Multiplying and Dividing Mixed Numbers

✎ **Find the product.**

1) $5\frac{1}{2} \times 2\frac{1}{4} =$

2) $5\frac{1}{3} \times 4\frac{1}{3} =$

3) $5\frac{3}{4} \times 6\frac{1}{4} =$

4) $3\frac{1}{3} \times 2\frac{3}{5} =$

5) $4\frac{8}{10} \times 1\frac{1}{24} =$

6) $6\frac{2}{7} \times 1\frac{1}{11} =$

7) $8\frac{2}{3} \times 3\frac{1}{2} =$

8) $3\frac{4}{7} \times 2\frac{1}{5} =$

9) $5\frac{2}{8} \times 4\frac{1}{6} =$

10) $7\frac{3}{3} \times 1\frac{3}{8} =$

✎ **Find the quotient.**

11) $2\frac{2}{5} \div 4\frac{1}{5} =$

12) $4\frac{1}{6} \div 3\frac{1}{3} =$

13) $6\frac{1}{3} \div 1\frac{1}{2} =$

14) $7\frac{1}{10} \div 2\frac{2}{5} =$

15) $3\frac{1}{3} \div 1\frac{1}{9} =$

16) $1\frac{1}{10} \div 4\frac{1}{2} =$

17) $1\frac{3}{16} \div 5\frac{1}{4} =$

18) $4\frac{1}{3} \div 4\frac{3}{4} =$

19) $9\frac{1}{3} \div 2\frac{1}{4} =$

20) $15\frac{1}{3} \div 5\frac{1}{2} =$

21) $4\frac{1}{6} \div 1\frac{1}{5} =$

22) $1\frac{1}{18} \div 1\frac{2}{9} =$

23) $4\frac{2}{7} \div 1\frac{3}{10} =$

24) $7\frac{1}{3} \div 2\frac{2}{11} =$

25) $8\frac{2}{5} \div 1\frac{1}{6} =$

26) $9\frac{1}{3} \div 2\frac{1}{7} =$

Adding and Subtracting Decimals

✎ **Add and subtract decimals.**

$$
\begin{array}{r}
1)\quad \begin{array}{r} 35.19 \\ -\ 24.28 \\ \hline \end{array}
\end{array}
\qquad
4)\quad \begin{array}{r} 38.72 \\ -\ 21.68 \\ \hline \end{array}
\qquad
7)\quad \begin{array}{r} 86.09 \\ -\ 35.14 \\ \hline \end{array}
$$

$$
2)\quad \begin{array}{r} 34.29 \\ +\ 42.58 \\ \hline \end{array}
\qquad
5)\quad \begin{array}{r} 57.39 \\ +\ 26.54 \\ \hline \end{array}
\qquad
8)\quad \begin{array}{r} 54.51 \\ +\ 32.66 \\ \hline \end{array}
$$

$$
3)\quad \begin{array}{r} 61.20 \\ +\ 33.75 \\ \hline \end{array}
\qquad
6)\quad \begin{array}{r} 70.24 \\ -\ 42.35 \\ \hline \end{array}
\qquad
9)\quad \begin{array}{r} 114.21 \\ -\ 88.69 \\ \hline \end{array}
$$

✎ **Find the missing number.**

10) ___ $+\ 2.8 = 5.4$

11) $4.1 +$ ___ $= 5.88$

12) $6.45 +$ ___ $= 8$

13) $7.25 -$ ___ $= 3.40$

14) ___ $-\ 2.35 = 4.25$

15) ___ $-\ 19.85 = 6.54$

16) $22.15 +$ ___ $= 28.95$

17) ___ $-\ 37.16 = 9.42$

18) ___ $+\ 24.50 = 34.19$

19) $72.40 +$ ___ $= 125.20$

Multiplying and Dividing Decimals

✎ Find the product.

1) $0.5 \times 0.6 =$

2) $3.3 \times 0.4 =$

3) $1.28 \times 0.5 =$

4) $0.35 \times 0.6 =$

5) $1.85 \times 0.6 =$

6) $0.24 \times 0.5 =$

7) $5.25 \times 1.4 =$

8) $18.5 \times 4.6 =$

9) $15.4 \times 6.8 =$

10) $19.5 \times 2.6 =$

11) $32.2 \times 1.5 =$

12) $78.4 \times 4.5 =$

✎ Find the quotient.

13) $1.85 \div 10 =$

14) $74.6 \div 100 =$

15) $3.6 \div 3 =$

16) $9.6 \div 0.4 =$

17) $15.5 \div 0.5 =$

18) $32.8 \div 0.2 =$

19) $22.15 \div 1,000 =$

20) $53.55 \div 0.7 =$

21) $322.2 \div 0.2 =$

22) $50.67 \div 0.18 =$

23) $77.4 \div 0.8 =$

24) $27.93 \div 0.03 =$

Comparing Decimals

✎ **Write the correct comparison symbol (>, < or =).**

1) 0.70 ☐ 0.070

2) 0.049 ☐ 0.49

3) 5.090 ☐ 5.09

4) 2.57 ☐ 2.05

5) 9.03 ☐ 0.930

6) 6.06 ☐ 6.6

7) 7.02 ☐ 7.020

8) 3.04 ☐ 3.2

9) 3.61 ☐ 3.245

10) 0.986 ☐ 0.0986

11) 17.24 ☐ 17.240

12) 0.759 ☐ 0.81

13) 9.040 ☐ 9.40

14) 5.73 ☐ 5.213

15) 9.44 ☐ 9.404

16) 7.17 ☐ 7.170

17) 4.85 ☐ 4.085

18) 9.041 ☐ 9.40

19) 3.033 ☐ 3.030

20) 4.97 ☐ 4.970

Rounding Decimals

✍ **Round each decimal to the nearest whole number.**

1) 28.12	3) 16.22	5) 7.95
2) 6.9	4) 8.5	6) 52.7

✍ **Round each decimal to the nearest tenth.**

7) 31.761	9) 94.729	11) 13.219
8) 14.421	10) 77.89	12) 59.89

✍ **Round each decimal to the nearest hundredth.**

13) 8.428	15) 55.3786	17) 62.241
14) 23.812	16) 231.912	18) 19.447

✍ **Round each decimal to the nearest thousandth.**

19) 15.54324	21) 243.8652	23) 67.1983
20) 34.62586	22) 80.4529	24) 72.36788

Answers of Worksheets

Simplifying Fractions

1) $\frac{1}{2}$

2) $\frac{4}{5}$

3) $\frac{3}{4}$

4) $\frac{1}{2}$

5) $\frac{1}{4}$

6) $\frac{2}{3}$

7) $\frac{4}{5}$

8) $\frac{1}{4}$

9) $\frac{1}{2}$

10) $\frac{5}{7}$

11) $\frac{1}{4}$

12) $\frac{1}{2}$

13) $\frac{7}{8}$

14) $\frac{9}{10}$

15) $\frac{1}{3}$

16) $\frac{5}{14}$

17) $\frac{2}{7}$

18) $\frac{4}{27}$

19) $\frac{6}{31}$

20) $\frac{4}{9}$

21) $\frac{1}{8}$

22) C

23) A

24) B

Adding and Subtracting Fractions

1) $\frac{9}{9} = 1$

2) $\frac{9}{14}$

3) $\frac{5}{8}$

4) $1\frac{1}{10}$

5) $\frac{17}{20}$

6) $1\frac{1}{4}$

7) $1\frac{1}{5}$

8) $1\frac{1}{15}$

9) $1\frac{8}{21}$

10) $1\frac{1}{3}$

11) $1\frac{7}{30}$

12) $\frac{3}{4}$

13) $\frac{1}{6}$

14) $\frac{5}{8}$

15) $\frac{1}{6}$

16) $\frac{1}{20}$

17) $-\frac{1}{24}$

18) $\frac{3}{28}$

19) $\frac{13}{18}$

20) $\frac{7}{12}$

21) $\frac{19}{24}$

22) $-\frac{1}{15}$

23) $\frac{5}{28}$

24) $\frac{1}{2}$

25) $\frac{3}{28}$

26) $\frac{27}{40}$

27) $\frac{18}{35}$

28) $\frac{5}{16}$

29) $\frac{1}{2}$

30) $\frac{1}{18}$

Multiplying and Dividing Fractions

1) $\frac{3}{5}$

2) $\frac{1}{3}$

3) $\frac{1}{20}$

4) $\frac{1}{30}$

5) $\frac{1}{20}$

6) $\frac{1}{9}$

7) $\frac{2}{7}$

8) $\frac{1}{16}$

9) $\frac{3}{8}$

10) $\frac{1}{14}$

11) $\frac{1}{3}$

12) $\frac{1}{6}$

13) 2

14) $\frac{1}{2}$

15) $3\frac{3}{4}$

16) $\frac{2}{5}$

17) $\frac{3}{4}$

18) $4\frac{1}{2}$

19) $\frac{26}{49}$

20) $\frac{2}{9}$

21) $\frac{7}{10}$

22) 1

23) $\frac{3}{5}$

24) $\frac{3}{4}$

25) $\frac{5}{18}$

26) $\frac{25}{36}$

27) $\frac{9}{10}$

28) $\frac{8}{15}$

29) $3\frac{3}{4}$

30) $\frac{20}{33}$

Adding and Subtracting Mixed Numbers

1) $5\frac{1}{2}$

2) 8

3) $4\frac{1}{2}$

4) $4\frac{7}{12}$

5) $6\frac{5}{12}$

6) $8\frac{13}{15}$

7) $6\frac{16}{21}$

8) $7\frac{9}{10}$

9) $11\frac{29}{35}$

10) $14\frac{19}{48}$

11) $1\frac{1}{2}$

12) $2\frac{1}{5}$

13) $1\frac{2}{9}$

14) $1\frac{9}{14}$

15) $3\frac{1}{4}$

16) $3\frac{13}{15}$

17) $3\frac{1}{8}$

18) $2\frac{13}{15}$

19) $3\frac{7}{30}$

20) $4\frac{3}{35}$

21) $5\frac{1}{12}$

22) $2\frac{1}{16}$

23) $4\frac{1}{24}$

24) $2\frac{71}{72}$

25) $5\frac{33}{35}$

26) $6\frac{32}{63}$

Multiplying and Dividing Mixed Numbers

1) $12\frac{3}{8}$

2) $23\frac{1}{9}$

3) $35\frac{15}{16}$

4) $8\frac{2}{3}$

5) 5

6) $6\frac{6}{7}$

7) $30\frac{1}{3}$

8) $7\frac{6}{7}$

9) $21\frac{7}{8}$

10) 11

11) $\frac{4}{7}$

12) $1\frac{1}{4}$

13) $4\frac{2}{9}$

14) $2\frac{23}{24}$

15) 3

16) $\frac{11}{45}$

17) $\frac{19}{84}$

18) $\frac{52}{57}$

19) $4\frac{4}{27}$

20) $2\frac{26}{33}$

21) $3\frac{17}{36}$

22) $\frac{19}{22}$

23) $3\frac{27}{91}$

24) $3\frac{13}{36}$

25) $7\frac{1}{5}$

26) $4\frac{16}{45}$

Adding and Subtracting Decimals

1) 10.91

2) 76.87

3) 94.95

4) 17.04

5) 83.93   9) 25.52   13) 3.85   17) 46.58
6) 27.89   10) 2.6   14) 6.6   18) 9.69
7) 50.95   11) 1.78   15) 26.39   19) 52.8
8) 87.17   12) 1.55   16) 6.8

Multiplying and Dividing Decimals

1) 0.3   7) 7.35   13) 0.185   19) 0.02215
2) 1.32   8) 85.1   14) 0.746   20) 76.5
3) 0.64   9) 104.72   15) 1.2   21) 1,611
4) 0.21   10) 50.7   16) 24   22) 281.5
5) 1.11   11) 48.3   17) 31   23) 96.75
6) 0.12   12) 352.8   18) 164   24) 931

Comparing Decimals

1) >   6) <   11) =   16) =
2) <   7) =   12) <   17) >
3) =   8) <   13) <   18) <
4) >   9) >   14) >   19) >
5) >   10) >   15) >   20) =

Rounding Decimals

1) 28   9) 94.7   17) 62.24
2) 7   10) 77.9   18) 19.45
3) 16   11) 13.2   19) 15.543
4) 9   12) 59.9   20) 34.626
5) 8   13) 8.43   21) 243.865
6) 53   14) 23.81   22) 80.453
7) 31.8   15) 55.38   23) 67.198
8) 14.4   16) 231.91 24) 72.368

Chapter 3 :

Proportions, Ratios, and Percent

Topics that you will practice in this chapter:

- ✓ Simplifying Ratios
- ✓ Proportional Ratios
- ✓ Similarity and Ratios
- ✓ Ratio and Rates Word Problems
- ✓ Percentage Calculations
- ✓ Percent Problems
- ✓ Discount, Tax and Tip
- ✓ Percent of Change
- ✓ Simple Interest

Without mathematics, there's nothing you can do. Everything around you is mathematics. Everything around you is numbers." – Shakuntala Devi

Simplifying Ratios

✍ **Reduce each ratio.**

1) $15 : 20 = $ ___ : ___

2) $7 : 70 = $ ___ : ___

3) $16 : 28 = $ ___ : ___

4) $7 : 21 = $ ___ : ___

5) $4 : 40 = $ ___ : ___

6) $6 : 48 = $ ___ : ___

7) $16 : 64 = $ ___ : ___

8) $10 : 25 = $ ___ : ___

9) $8 : 48 = $ ___ : ___

10) $49 : 63 = $ ___ : ___

11) $18 : 27 = $ ___ : ___

12) $35 : 10 = $ ___ : ___

13) $90 : 9 = $ ___ : ___

14) $24 : 32 = $ ___ : ___

15) $7 : 56 = $ ___ : ___

16) $45 : 63 = $ ___ : ___

17) $56 : 72 = $ ___ : ___

18) $26 : 13 = $ ___ : ___

19) $15 : 45 = $ ___ : ___

20) $28 : 4 = $ ___ : ___

21) $24 : 48 = $ ___ : ___

22) $30 : 24 = $ ___ : ___

23) $70 : 140 = $ ___ : ___

24) $6 : 180 = $ ___ : ___

✍ **Write each ratio as a fraction in simplest form.**

25) $6 : 12 = $

26) $30 : 50 = $

27) $15 : 35 = $

28) $9 : 27 = $

29) $8 : 24 = $

30) $18 : 84 = $

31) $7 : 14 = $

32) $7 : 35 = $

33) $40 : 96 = $

34) $12 : 54 = $

35) $44 : 52 = $

36) $12 : 27 = $

37) $15 : 180 = $

38) $39 : 143 = $

39) $20 : 300 = $

40) $30 : 120 = $

41) $56 : 42 = $

42) $26 : 130 = $

43) $66 : 123 = $

44) $70 : 630 = $

45) $75 : 125 = $

Proportional Ratios

✍ **Fill in the blanks; Calculate each proportion.**

1) $3:8 = \underline{\quad} : 48$

2) $2:5 = 20:\underline{\quad}$

3) $1:9 = \underline{\quad} : 81$

4) $6:7 = 12:\underline{\quad}$

5) $9:2 = 63:\underline{\quad}$

6) $8:7 = \underline{\quad} : 49$

7) $20:3 = \underline{\quad} : 15$

8) $1:3 = \underline{\quad} : 75$

9) $7:6 = \underline{\quad} : 60$

10) $8:5 = \underline{\quad} : 45$

11) $3:10 = 60:\underline{\quad}$

12) $6:11 = 42:\underline{\quad}$

✍ **State if each pair of ratios form a proportion.**

13) $\frac{3}{20}$ and $\frac{9}{60}$

14) $\frac{1}{7}$ and $\frac{6}{42}$

15) $\frac{3}{7}$ and $\frac{24}{56}$

16) $\frac{4}{9}$ and $\frac{12}{18}$

17) $\frac{1}{9}$ and $\frac{12}{81}$

18) $\frac{7}{8}$ and $\frac{21}{28}$

19) $\frac{9}{13}$ and $\frac{27}{39}$

20) $\frac{1}{8}$ and $\frac{8}{64}$

21) $\frac{6}{19}$ and $\frac{30}{85}$

22) $\frac{5}{9}$ and $\frac{40}{81}$

23) $\frac{9}{14}$ and $\frac{108}{168}$

24) $\frac{15}{23}$ and $\frac{360}{552}$

✍ **Calculate each proportion.**

25) $\frac{20}{25} = \frac{32}{x}, x = \underline{\quad}$

26) $\frac{1}{8} = \frac{32}{x}, x = \underline{\quad}$

27) $\frac{15}{5} = \frac{21}{x}, x = \underline{\quad}$

28) $\frac{1}{7} = \frac{x}{294}, x = \underline{\quad}$

29) $\frac{7}{9} = \frac{x}{81}, x = \underline{\quad}$

30) $\frac{1}{5} = \frac{13}{x}, x = \underline{\quad}$

31) $\frac{9}{5} = \frac{36}{x}, x = \underline{\quad}$

32) $\frac{6}{13} = \frac{48}{x}, x = \underline{\quad}$

33) $\frac{5}{8} = \frac{x}{88}, x = \underline{\quad}$

34) $\frac{4}{15} = \frac{x}{240}, x = \underline{\quad}$

35) $\frac{9}{19} = \frac{x}{266}, x = \underline{\quad}$

36) $\frac{7}{15} = \frac{x}{270}, x = \underline{\quad}$

Similarity and Ratios

✎ **Each pair of figures is similar. Find the missing side.**

1)
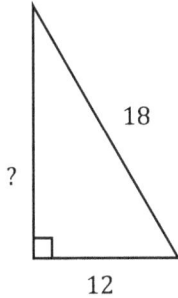

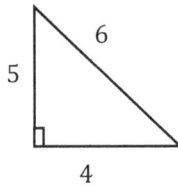

2)
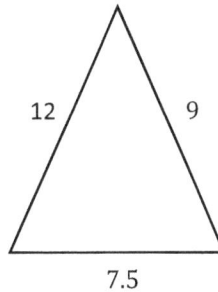

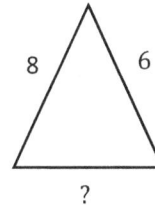

3)
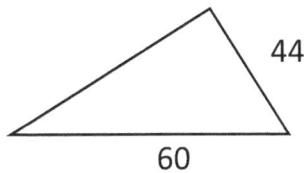

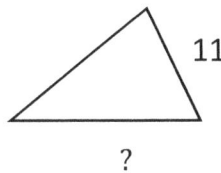

4)
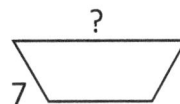

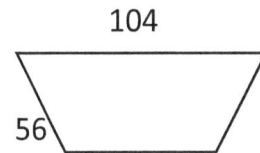

✎ **Calculate.**

5) Two rectangles are similar. The first is 24 feet wide and 120 feet long. The second is 30 feet wide. What is the length of the second rectangle?
_____

6) Two rectangles are similar. One is 5 meters by 36 meters. The longer side of the second rectangle is 90 meters. What is the other side of the second rectangle? _____

7) A building casts a shadow 25 ft long. At the same time a girl 10 ft tall casts a shadow 5 ft long. How tall is the building? _____

8) The scale of a map of Texas is 4 inches: 32 miles. If you measure the distance from Dallas to Martin County as 38.4 inches, approximately how far is Martin County from Dallas? _____

Ratio and Rates Word Problems

✍ **Find the answer for each word problem.**

1) Mason has 24 red cards and 36 green cards. What is the ratio of Mason 's red cards to his green cards? _____

2) In a party, 45 soft drinks are required for every 54 guests. If there are 378 guests, how many soft drinks is required? _____

3) In Mason's class, 42 of the students are tall and 24 are short. In Michael's class 84 students are tall and 48 students are short. Which class has a higher ratio of tall to short students? _____

4) The price of 5 apples at the Quick Market is $4.6. The price of 7 of the same apples at Walmart is $5.95. Which place is the better buy? _____

5) The bakers at a Bakery can make 90 bagels in 3 hours. How many bagels can they bake in 24 hours? What is that rate per hour? _____

6) You can buy 5 cans of green beans at a supermarket for $5.75. How much does it cost to buy 45 cans of green beans? _____

7) The ratio of boys to girls in a class is 4: 7. If there are 32 boys in the class, how many girls are in that class? _____

8) The ratio of red marbles to blue marbles in a bag is 3: 7. If there are 50 marbles in the bag, how many of the marbles are red? _____

Percentage Calculations

✎ **Calculate the given percent of each value.**

1) 3% of 60 = ___

2) 20% of 32 = ___

3) 4% of 72 = ___

4) 16% of 32 = ___

5) 25% of 124 = ___

6) 35% of 56 = ___

7) 15% of 20 = ___

8) 14% of 150 = ___

9) 80% of 50 = ___

10) 12% of 115 = ___

11) 72% of 250 = ___

12) 52% of 500 = ___

13) 70% of 400 = ___

14) 27% of 145 = ___

15) 90% of 64 = ___

16) 60% of 55 = ___

17) 22% of 210 = ___

18) 8% of 235 = ___

✎ **Calculate the percent of each given value.**

19) ___% of 25 = 5

20) ___% of 40 = 20

21) ___% of 25 = 2

22) ___% of 50 = 16

23) ___% of 250 = 5

24) ___% of 40 = 32

25) ___% of 125 = 20

26) ___% of 700 = 49

27) ___% of 350 = 49

28) ___% of 500 = 210

✎ **Calculate each percent problem.**

29) A Cinema has 250 seats. 60 seats were sold for the current movie. What percent of seats are empty? _____ %

30) There are 68 boys and 92 girls in a class. 75% of the students in the class take the bus to school. How many students do not take the bus to school? _____

Percent Problems

✎ **Calculate each problem.**

1) 9 is what percent of 45? ____%

2) 60 is what percent of 120? ____%

3) 10 is what percent of 200? ____%

4) 15 is what percent of 125? ____%

5) 10 is what percent of 400? ____%

6) 66 is what percent of 55? ____%

7) 40 is what percent of 160? ____%

8) 40 is what percent of 50? ____%

9) 120 is what percent of 800? ____%

10) 78 is what percent of 120? ___%

11) 36 is what percent of 144? ___%

12) 17 is what percent of 85? ___%

13) 90 is what percent of 900? ___%

14) 36 is what percent of 16? ___%

15) 63 is what percent of 14? ___%

16) 18 is what percent of 60? ___%

17) 126 is what percent of 200? ___%

18) 232 is what percent of 40? ___%

✎ **Calculate each percent word problem.**

19) There are 40 employees in a company. On a certain day, 25 were present. What percent showed up for work? ____%

20) A metal bar weighs 60 ounces. 25% of the bar is gold. How many ounces of gold are in the bar? _____

21) A crew is made up of 12 women; the rest are men. If 15% of the crew are women, how many people are in the crew? _____

22) There are 40 students in a class and 8 of them are girls. What percent are boys? ____%

23) The Royals softball team played 400 games and won 280 of them. What percent of the games did they lose? ____%

Discount, Tax and Tip

✎ Find the selling price of each item.

1) Original price of a computer: $420

Tax: 8% Selling price: $_____

2) Original price of a laptop: $280

Tax: 4% Selling price: $_____

3) Original price of a sofa: $820

Tax: 5% Selling price: $_____

4) Original price of a car: $15,800

Tax: 3.6% Selling price: $_____

5) Original price of a Table: $250

Tax: 9% Selling price: $_____

6) Original price of a house: $630,000

Tax: 1.8% Selling price: $_____

7) Original price of a tablet: $450

Discount: 30% Selling price: $____

8) Original price of a chair: $390

Discount: 8% Selling price: $____

9) Original price of a book: $75

Discount: 42% Selling price: $____

10) Original price of a cellphone: $820

Discount: 23% Selling price: $___

11) Food bill: $45

Tip: 15% Price: $_____

12) Food bill: $32

Tipp: 20% Price: $_____

13) Food bill: $90

Tip: 35% Price: $_____

14) Food bill: $42

Tipp: 12% Price: $_____

✎ Find the answer for each word problem.

15) Nicolas hired a moving company. The company charged $500 for its services, and Nicolas gives the movers a 40% tip. How much does Nicolas tip the movers? $_____

16) Mason has lunch at a restaurant and the cost of his meal is $90. Mason wants to leave a 25% tip. What is Mason's total bill including tip? $_____

17) The sales tax in Texas is 19.80% and an item costs $350. How much is the tax? $_____

18) The price of a table at Best Buy is $680. If the sales tax is 5%, what is the final price of the table including tax? $_____

Percent of Change

✎ Find each percent of change.

1) From 150 to 450. ___ %

2) From 50 ft to 250 ft. ___ %

3) From $60 to $360. ___ %

4) From 60 cm to 180 cm. ___ %

5) From 15 to 45. ___ %

6) From 80 to 16. ___ %

7) From 120 to 360. ___ %

8) From 900 to 450. ___ %

9) From 1,000 to 200. ___ %

10) From 144 to 36. ___ %

✎ Calculate each percent of change word problem.

11) Bob got a raise, and his hourly wage increased from $42 to $63. What is the percent increase? ___ %

12) The price of a pair of shoes increases from $50 to $61. What is the percent increase? ___ %

13) At a coffee shop, the price of a cup of coffee increased from $4.80 to $5.76. What is the percent increase in the cost of the coffee? ___ %

14) 51 cm are cut from 85 cm board. What is the percent decrease in length? ___ %

15) In a class, the number of students has been increased from 54 to 81. What is the percent increase? ___ %

16) The price of gasoline rises from $24.40 to $30.50 in one month. By what percent did the gas price rise? ___ %

17) A shirt was originally priced at $38. It went on sale for $24.70. What was the percent that the shirt was discounted? ___ %

Simple Interest

✍ **Determine the simple interest for these loans.**

1) $480 at 11% for 3 years. $ _____

2) $4,200 at 7% for 4 years. $ _____

3) $2,500 at 20% for 3 years. $ _____

4) $6,800 at 3.9% for 4 months. $ ____

5) $800 at 6% for 7 months. $ _____

6) $36,000 at 4.2% for 6 years. $ _____

7) $6,500 at 7% for 4 years. $ _____

8) $850 at 9.5% for 2 years. $ _____

9) $1,200 at 5.8% for 9 months. $ ____

10) $3,000 at 4.5% for 7 years. $ _____

✍ **Calculate each simple interest word problem.**

11) A new car, valued at $22,000, depreciates at 8.5% per year. What is the value of the car one year after purchase? $_____

12) Sara puts $9,000 into an investment yielding 6% annual simple interest; she left the money in for three years. How much interest does Sara get at the end of those three years? $_____

13) A bank is offering 12% simple interest on a savings account. If you deposit $16,400, how much interest will you earn in two years? $_____

14) $720 interest is earned on a principal of $6,000 at a simple interest rate of 4% interest per year. For how many years was the principal invested? _____

15) In how many years will $2,200 yield an interest of $440 at 4% simple interest? _____

16) Jim invested $8,000 in a bond at a yearly rate of 4.5%. He earned $1,440 in interest. How long was the money invested? _____

Answers of Worksheets

Simplifying Ratios

1) $3:4$	14) $3:4$	26) $\frac{3}{5}$	36) $\frac{4}{9}$
2) $1:10$	15) $1:8$	27) $\frac{3}{7}$	37) $\frac{1}{12}$
3) $4:7$	16) $5:7$	28) $\frac{1}{3}$	38) $\frac{3}{11}$
4) $1:3$	17) $7:9$	29) $\frac{1}{3}$	39) $\frac{1}{15}$
5) $1:10$	18) $2:1$	30) $\frac{3}{14}$	40) $\frac{1}{4}$
6) $1:8$	19) $1:3$	31) $\frac{1}{2}$	41) $\frac{4}{3}$
7) $2:8$	20) $7:1$	32) $\frac{1}{5}$	42) $\frac{1}{5}$
8) $2:5$	21) $1:2$	33) $\frac{5}{12}$	43) $\frac{22}{41}$
9) $1:6$	22) $5:4$	34) $\frac{2}{9}$	44) $\frac{1}{9}$
10) $7:9$	23) $1:2$	35) $\frac{11}{13}$	45) $\frac{3}{5}$
11) $2:3$	24) $1:30$		
12) $7:2$	25) $\frac{1}{2}$		
13) $10:1$			

Proportional Ratios

1) 18	10) 72	19) Yes	28) 42
2) 50	11) 200	20) Yes	29) 63
3) 9	12) 77	21) No	30) 65
4) 14	13) Yes	22) No	31) 20
5) 14	14) Yes	23) Yes	32) 104
6) 56	15) Yes	24) Yes	33) 55
7) 100	16) No	25) 40	34) 64
8) 25	17) No	26) 256	35) 126
9) 70	18) No	27) 7	36) 126

Similarity and ratios

1) 15	4) 13	7) 50 feet
2) 5	5) 150 feet	8) 307.2 miles
3) 15	6) 12.5 meters	

Ratio and Rates Word Problems

1) $2:3$	2) 315

3) The ratio for both classes is 7 to 4.

4) Walmart is a better buy.

5) 720, the rate is 30 per hour.

6) $51.75

7) 56

8) 15

Percentage Calculations

1) 1.8

2) 6.4

3) 2.88

4) 5.12

5) 31

6) 19.6

7) 3

8) 21

9) 40

10) 13.8

11) 180

12) 260

13) 280

14) 39.15

15) 57.6

16) 33

17) 46.2

18) 18.8

19) 20%

20) 50%

21) 8%

22) 32%

23) 2%

24) 80%

25) 16%

26) 7%

27) 14%

28) 42%

29) 76%

30) 40

Percent Problems

1) 20%

2) 50%

3) 5%

4) 12%

5) 2.5%

6) 120%

7) 25%

8) 80%

9) 15%

10) 65%

11) 25%

12) 20%

13) 10%

14) 225%

15) 450%

16) 30%

17) 63%

18) 580%

19) 62.5%

20) 15 ounces

21) 80

22) 80%

23) 30%

Discount, Tax and Tip

1) $453.60

2) $291.20

3) $861.00

4) $16,368.80

5) $272.50

6) $641,340

7) $315.00

8) $358.80

9) $43.50

10) $631.40

11) $51.75

12) $38.40

13) $121.50

14) $47.04

15) $200.00

16) $112.50

17) $69.30

18) $714.00

Percent of Change

1) 200%	7) 200%	13) 20%
2) 400%	8) 50%	14) 60%
3) 500%	9) 80%	15) 50%
4) 200%	10) 75%	16) 25%
5) 200%	11) 50%	17) 35%
6) 80%	12) 22%	

Simple Interest

1) $158.40	7) $1,820.00	13) $3,936.00
2) $1,176.00	8) $161.50	14) 3 years
3) $1,500.00	9) $52.20	15) 5 years
4) $88.40	10) $945.00	16) 4 years
5) $28.00	11) $20,130.00	
6) $9,072.00	12) $1,620.00	

Chapter 4 :

Exponents and Radicals Expressions

Topics that you will practice in this chapter:

- ✓ Multiplication Property of Exponents
- ✓ Zero and Negative Exponents
- ✓ Division Property of Exponents
- ✓ Powers of Products and Quotients
- ✓ Negative Exponents and Negative Bases
- ✓ Scientific Notation
- ✓ Square Roots
- ✓ Simplifying Radical Expressions
- ✓ Simplifying Radical Expressions Involving Fractions
- ✓ Multiplying Radical Expressions
- ✓ Adding and Subtracting Radical Expressions

Mathematics is no more computation than typing is literature.

– John Allen Paulos

Multiplication Property of Exponents

✎ **Simplify and write the answer in exponential form.**

1) $4 \times 4^5 =$

2) $8^4 \times 8 =$

3) $7^3 \times 7^3 =$

4) $9^2 \times 9^2 =$

5) $2^2 \times 2^4 \times 2 =$

6) $5 \times 5^3 \times 5^3 =$

7) $4^3 \times 4^2 \times 4 \times 4 =$

8) $5x \times x =$

9) $x^3 \times x^3 =$

10) $x^7 \times x^2 =$

11) $x^4 \times x^3 \times x^2 =$

12) $10x \times 3x =$

13) $4x^3 \times 4x^3 =$

14) $7x^3 \times x =$

15) $3x^2 \times 4x^2 \times x^2 =$

16) $5x^4 \times x^4 =$

17) $2x^8 \times 2x =$

18) $6x \times x^5 =$

19) $4x^2 \times 6x^6 =$

20) $5yx^3 \times 4x =$

21) $7x^3 \times y^5x^7 =$

22) $y^2x^3 \times y^5x^4 =$

23) $3x^5 \times 4x^3y^4 =$

24) $4x^4 \times 9x^2y^5 =$

25) $5x^3y^4 \times 6x^8y^2 =$

26) $8x^3y^6 \times 4xy^3 =$

27) $2xy^5 \times 6x^3y^3 =$

28) $4x^5y^2 \times 4x^2y^8 =$

29) $7x \times 3y^8x^2 \times y^5 =$

30) $x^3 \times 2y^3x^4 \times 2y =$

31) $3yx^4 \times 3y^4x \times 3xy^3 =$

32) $6y^3 \times 2y^2x^4 \times 10yx^5 =$

Zero and Negative Exponents

✎ **Evaluate the following expressions.**

1) $1^{-5} =$

2) $4^{-1} =$

3) $0^{10} =$

4) $1^{15} =$

5) $5^{-2} =$

6) $3^{-3} =$

7) $9^{-1} =$

8) $10^{-2} =$

9) $12^{-2} =$

10) $2^{-5} =$

11) $3^{-4} =$

12) $2^{-4} =$

13) $6^{-3} =$

14) $10^{-3} =$

15) $30^{-1} =$

16) $15^{-2} =$

17) $4^{-3} =$

18) $2^{-7} =$

19) $5^{-3} =$

20) $4^{-4} =$

21) $3^{-5} =$

22) $10^{-4} =$

23) $2^{-10} =$

24) $8^{-3} =$

25) $20^{-2} =$

26) $14^{-2} =$

27) $9^{-3} =$

28) $100^{-2} =$

29) $5^{-4} =$

30) $4^{-6} =$

31) $\left(\frac{1}{4}\right)^{-3}$

32) $\left(\frac{1}{6}\right)^{-2} =$

33) $\left(\frac{1}{7}\right)^{-2} =$

34) $\left(\frac{2}{3}\right)^{-3} =$

35) $\left(\frac{1}{13}\right)^{-2} =$

36) $\left(\frac{7}{12}\right)^{-2} =$

37) $\left(\frac{1}{6}\right)^{-3} =$

38) $\left(\frac{1}{300}\right)^{-2} =$

39) $\left(\frac{2}{9}\right)^{-2} =$

40) $\left(\frac{7}{5}\right)^{-1} =$

41) $\left(\frac{13}{23}\right)^{0} =$

42) $\left(\frac{1}{4}\right)^{-5} =$

Division Property of Exponents

✎ **Simplify.**

1) $\dfrac{5^6}{5^7} =$

2) $\dfrac{8^8}{8^6} =$

3) $\dfrac{4^5}{4} =$

4) $\dfrac{3}{3^5} =$

5) $\dfrac{x}{x^6} =$

6) $\dfrac{3 \times 3^2}{3^2 \times 3^5} =$

7) $\dfrac{9^4}{9^2} =$

8) $\dfrac{10 \times 10^9}{10^2 \times 10^7} =$

9) $\dfrac{7^5 \times 7^7}{7^4 \times 7^8} =$

10) $\dfrac{15x}{30x^6} =$

11) $\dfrac{3x^9}{4x^4} =$

12) $\dfrac{15x^8}{10x^9} =$

13) $\dfrac{42x^5}{6y^9} =$

14) $\dfrac{36y^8}{4x^4y^5} =$

15) $\dfrac{2x^7}{9x} =$

16) $\dfrac{49x^8y^6}{7x^9} =$

17) $\dfrac{48x^2}{24x^6y^{12}} =$

18) $\dfrac{30yx^5}{6yx^7} =$

19) $\dfrac{19x^7y}{38x^{12}y^4} =$

20) $\dfrac{9x^8}{63x^8} =$

21) $\dfrac{9x^{-9}}{4x^{-3}} =$

Powers of Products and Quotients

✎ **Simplify.**

1) $(4^3)^2 =$

2) $(2^3)^4 =$

3) $(2 \times 2^3)^2 =$

4) $(5 \times 5^5)^6 =$

5) $(19^4 \times 19^2)^3 =$

6) $(2^3 \times 2^4)^4 =$

7) $(5 \times 5^2)^2 =$

8) $(4^4)^4 =$

9) $(8x^5)^2 =$

10) $(3x^2 y^4)^4 =$

11) $(7x^5 y^2)^2 =$

12) $(5x^4 y^4)^3 =$

13) $(2x^3 y^3)^5 =$

14) $(10x^3 y^4)^3 =$

15) $(13y^3 y)^2 =$

16) $(5x^6 x^4)^2 =$

17) $(6x^7 y^6)^3 =$

18) $(12x^5 x^7)^2 =$

19) $(2x^4 \times 2x)^4 =$

20) $(2x^4 y^3)^5 =$

21) $(15x^7 y^2)^2 =$

22) $(8x^3 y^5)^3 =$

23) $(3x \times 2y^2)^4 =$

24) $\left(\frac{4x}{x^5}\right)^2 =$

25) $\left(\frac{x^4 y^5}{x^3 y^5}\right)^9 =$

26) $\left(\frac{36xy}{6x^5}\right)^3 =$

27) $\left(\frac{x^7}{x^8 y^2}\right)^6 =$

28) $\left(\frac{xy^4}{x^3 y^6}\right)^{-3} =$

29) $\left(\frac{5xy^8}{x^3}\right)^2 =$

30) $\left(\frac{xy^6}{2xy^3}\right)^{-4} =$

Negative Exponents and Negative Bases

✎ **Simplify.**

1) $-9^{-1} =$

2) $-9^{-2} =$

3) $-2^{-5} =$

4) $-x^{-7} =$

5) $11x^{-1} =$

6) $-8x^{-3} =$

7) $-12x^{-5} =$

8) $-9x^{-8}y^{-6} =$

9) $32x^{-5}y^{-1} =$

10) $10a^{-9}b^{-3} =$

11) $-17x^4y^{-6} =$

12) $-\dfrac{25}{x^{-5}} =$

13) $-\dfrac{13x}{a^{-7}} =$

14) $\left(-\dfrac{1}{3}\right)^{-4} =$

15) $\left(-\dfrac{3}{4}\right)^{-2} =$

16) $-\dfrac{14}{a^{-6}b^{-3}} =$

17) $-\dfrac{7x}{x^{-8}} =$

18) $-\dfrac{a^{-9}}{b^{-5}} =$

19) $-\dfrac{11}{x^{-5}} =$

20) $\dfrac{8b}{-16c^{-6}} =$

21) $\dfrac{12ab}{a^{-4}b^{-3}} =$

22) $-\dfrac{8n^{-4}}{32p^{-7}} =$

23) $\dfrac{16ab^{-6}}{-6c^{-5}} =$

24) $\left(\dfrac{10a}{5c}\right)^{-4} =$

25) $\left(-\dfrac{12x}{4yz}\right)^{-3} =$

26) $\dfrac{8ab^{-7}}{-5c^{-3}} =$

27) $\left(-\dfrac{x^4}{x^5}\right)^{-5} =$

28) $\left(-\dfrac{x^{-2}}{7x^3}\right)^{-2} =$

29) $\left(-\dfrac{x^{-4}}{x^2}\right)^{-6} =$

Scientific Notation

✎ **Write each number in scientific notation.**

1) $0.223 =$

2) $0.09 =$

3) $4.5 =$

4) $900 =$

5) $2,000 =$

6) $0.006 =$

7) $33 =$

8) $9,400 =$

9) $1,470 =$

10) $52,000 =$

11) $8,000,000 =$

12) $0.00009 =$

13) $2,158,000 =$

14) $0.0039 =$

15) $0.000075 =$

16) $4,300,000 =$

17) $130,000 =$

18) $4,000,000,000 =$

19) $0.00009 =$

20) $0.0039 =$

✎ **Write each number in standard notation.**

21) $4 \times 10^{-1} =$

22) $1.2 \times 10^{-3} =$

23) $2.7 \times 10^{5} =$

24) $6 \times 10^{-4} =$

25) $3.6 \times 10^{-3} =$

26) $5.5 \times 10^{5} =$

27) $3.2 \times 10^{4} =$

28) $3.88 \times 10^{6} =$

29) $7 \times 10^{-6} =$

30) $4.2 \times 10^{-7} =$

Square Roots

✎ **Find the value each square root.**

1) $\sqrt{16} =$ ____

2) $\sqrt{25} =$ ____

3) $\sqrt{1} =$ ____

4) $\sqrt{64} =$ ____

5) $\sqrt{0} =$ ____

6) $\sqrt{196} =$ ____

7) $\sqrt{4} =$ ____

8) $\sqrt{256} =$ ____

9) $\sqrt{36} =$ ____

10) $\sqrt{289} =$ ____

11) $\sqrt{169} =$ ____

12) $\sqrt{144} =$ ____

13) $\sqrt{100} =$ ____

14) $\sqrt{1,600} =$ ____

15) $\sqrt{2,500} =$ ____

16) $\sqrt{324} =$ ____

17) $\sqrt{529} =$ ____

18) $\sqrt{20} =$ ____

19) $\sqrt{625} =$ ____

20) $\sqrt{18} =$ ____

21) $\sqrt{50} =$ ____

22) $\sqrt{1,024} =$ ____

23) $\sqrt{160} =$ ____

24) $\sqrt{32} =$ ____

✎ **Evaluate.**

25) $\sqrt{4} \times \sqrt{25} =$ _____

26) $\sqrt{36} \times \sqrt{49} =$ _____

27) $\sqrt{6} \times \sqrt{6} =$ _____

28) $\sqrt{13} \times \sqrt{13} =$ _____

29) $2\sqrt{5} \times 3\sqrt{5} =$ _____

30) $\sqrt{12} \times \sqrt{3} =$ _____

31) $\sqrt{13} + \sqrt{13} =$ _____

32) $\sqrt{10} + 2\sqrt{10} =$ _____

33) $12\sqrt{7} - 10\sqrt{7} =$ _____

34) $4\sqrt{10} \times 2\sqrt{10} =$ _____

35) $5\sqrt{3} \times 8\sqrt{3} =$ _____

36) $6\sqrt{3} - \sqrt{12} =$ _____

Simplifying Radical Expressions

✎ **Simplify.**

1) $\sqrt{13x^2} =$

2) $\sqrt{75x^2} =$

3) $\sqrt[3]{27a} =$

4) $\sqrt{64x^5} =$

5) $\sqrt{216a} =$

6) $\sqrt[3]{63w^3} =$

7) $\sqrt{192x} =$

8) $\sqrt{125v} =$

9) $\sqrt[3]{128x^2} =$

10) $\sqrt{100x^9} =$

11) $\sqrt{16x^4} =$

12) $\sqrt[3]{500a^5} =$

13) $\sqrt{242} =$

14) $\sqrt{392p^3} =$

15) $\sqrt{8m^6} =$

16) $\sqrt{198x^3y^3} =$

17) $\sqrt{121x^5y^5} =$

18) $\sqrt{16a^6b^3} =$

19) $\sqrt{90x^5y^7} =$

20) $\sqrt[3]{64y^2x^6} =$

21) $10\sqrt{16x^4} =$

22) $6\sqrt{81x^2} =$

23) $\sqrt[3]{56x^2y^6} =$

24) $\sqrt[3]{1,000x^5y^7} =$

25) $8\sqrt{50a} =$

26) $\sqrt[4]{625x^8y} =$

27) $\sqrt{24x^4y^5r^3} =$

28) $5\sqrt{36x^4y^5z^8} =$

29) $3\sqrt[3]{343x^9y^7} =$

30) $5\sqrt{81a^5b^2c^9} =$

31) $\sqrt[4]{625x^8y^{16}} =$

Multiplying Radical Expressions

✎ **Simplify.**

1) $\sqrt{5} \times \sqrt{5} =$

2) $\sqrt{5} \times \sqrt{10} =$

3) $\sqrt{3} \times \sqrt{12} =$

4) $\sqrt{49} \times \sqrt{47} =$

5) $\sqrt{7} \times -2\sqrt{28} =$

6) $3\sqrt{15} \times \sqrt{5} =$

7) $4\sqrt{72} \times \sqrt{2} =$

8) $\sqrt{5} \times -\sqrt{49} =$

9) $\sqrt{55} \times \sqrt{11} =$

10) $7\sqrt{42} \times 2\sqrt{216} =$

11) $\sqrt{45}(5 + \sqrt{5}) =$

12) $\sqrt{13x^2} \times \sqrt{13x^3} =$

13) $-2\sqrt{27} \times \sqrt{3} =$

14) $2\sqrt{13x^4} \times \sqrt{13x^4} =$

15) $\sqrt{14x^3} \times \sqrt{7x^2} =$

16) $-8\sqrt{5x} \times \sqrt{7x^5} =$

17) $-2\sqrt{16x^5} \times 4\sqrt{8x^3} =$

18) $-4\sqrt{32}(8 + \sqrt{32}) =$

19) $\sqrt{32x}(10 - \sqrt{2x}) =$

20) $\sqrt{2x}(8\sqrt{x^5} + \sqrt{8}) =$

21) $\sqrt{20r}(5 + \sqrt{5}) =$

22) $-4\sqrt{7x} \times 3\sqrt{14x^5} =$

23) $-2\sqrt{12x} \times 3\sqrt{2x}$

24) $-\sqrt{7v^3}(-3\sqrt{42v}) =$

25) $(\sqrt{11} - 5)(\sqrt{11} + 5) =$

26) $(-3\sqrt{5} + 3)(\sqrt{5} - 4) =$

27) $(4 - 6\sqrt{3})(-6 + \sqrt{3}) =$

28) $(8 - 3\sqrt{5})(7 - \sqrt{5}) =$

29) $(-1 - \sqrt{3x})(4 + \sqrt{3x}) =$

30) $(-5 + 2\sqrt{7r})(-5 + \sqrt{7r}) =$

31) $(-\sqrt{7n} + 1)(-\sqrt{7} - 5) =$

32) $(-3 + \sqrt{3})(5 - 2\sqrt{3x}) =$

Simplifying Radical Expressions Involving Fractions

✍ **Simplify.**

1) $\dfrac{\sqrt{5}}{\sqrt{3}} =$

2) $\dfrac{\sqrt{18}}{\sqrt{45}} =$

3) $\dfrac{\sqrt{10}}{5\sqrt{2}} =$

4) $\dfrac{13}{\sqrt{3}} =$

5) $\dfrac{12\sqrt{5r}}{\sqrt{m^5}} =$

6) $\dfrac{11\sqrt{2}}{\sqrt{k}} =$

7) $\dfrac{6\sqrt{20x^3}}{\sqrt{16x}} =$

8) $\dfrac{\sqrt{14x^3y^4}}{\sqrt{7x^4y^3}} =$

9) $\dfrac{1}{1-\sqrt{5}} =$

10) $\dfrac{1-8\sqrt{a}}{\sqrt{11a}} =$

11) $\dfrac{\sqrt{a}}{\sqrt{a}+\sqrt{b}} =$

12) $\dfrac{1-\sqrt{5}}{2-\sqrt{6}} =$

13) $\dfrac{4+\sqrt{7}}{3-\sqrt{8}} =$

14) $\dfrac{5}{-3-3\sqrt{3}} =$

15) $\dfrac{7}{2-\sqrt{5}} =$

16) $\dfrac{\sqrt{7}-\sqrt{3}}{\sqrt{3}-\sqrt{7}} =$

17) $\dfrac{\sqrt{5}+\sqrt{7}}{\sqrt{7}-\sqrt{5}} =$

18) $\dfrac{2\sqrt{2}-\sqrt{3}}{3\sqrt{2}+\sqrt{5}} =$

19) $\dfrac{\sqrt{11}+5\sqrt{3}}{4-\sqrt{11}} =$

20) $\dfrac{\sqrt{5}+\sqrt{3}}{2-\sqrt{3}} =$

21) $\dfrac{\sqrt{32a^7b^4}}{\sqrt{2ab^3}} =$

22) $\dfrac{10\sqrt{21x^5}}{5\sqrt{x^3}} =$

Adding and Subtracting Radical Expressions

✎ **Simplify.**

1) $\sqrt{2} + \sqrt{8} =$

2) $3\sqrt{50} + 4\sqrt{2} =$

3) $2\sqrt{12} - 4\sqrt{3} =$

4) $5\sqrt{32} - 5\sqrt{2} =$

5) $3\sqrt{75} - 5\sqrt{3} =$

6) $-\sqrt{72} - 4\sqrt{2} =$

7) $-7\sqrt{16} - 4\sqrt{25} =$

8) $8\sqrt{24} + 2\sqrt{6} =$

9) $10\sqrt{49} - 7\sqrt{100} =$

10) $-7\sqrt{5} + 9\sqrt{45} =$

11) $-15\sqrt{12} + 14\sqrt{48} =$

12) $20\sqrt{4} - 2\sqrt{25} =$

13) $-2\sqrt{20} + 7\sqrt{5} =$

14) $8\sqrt{7} - 2\sqrt{63} =$

15) $5\sqrt{44} + 3\sqrt{11} =$

16) $3\sqrt{27} - 5\sqrt{48} =$

17) $\sqrt{144} - \sqrt{81} =$

18) $3\sqrt{20} - 6\sqrt{5} =$

19) $-2\sqrt{7} + 8\sqrt{28} =$

20) $3\sqrt{75} - 2\sqrt{3} =$

21) $5\sqrt{27} - 3\sqrt{3} =$

22) $-7\sqrt{30} + 6\sqrt{120} =$

23) $-7\sqrt{24} - 2\sqrt{6} =$

24) $-\sqrt{32x} + 4\sqrt{2x} =$

25) $\sqrt{7y^2} + y\sqrt{112} =$

26) $\sqrt{45mn^2} + 2n\sqrt{5m} =$

27) $-4\sqrt{12a} - 4\sqrt{3a} =$

28) $-5\sqrt{15ab} - 2\sqrt{60ab} =$

29) $\sqrt{45x^2y} + x\sqrt{20y} =$

30) $2\sqrt{7a} + 4\sqrt{63a} =$

SSAT Upper-Level Subject Test Mathematics

Answers of Worksheets

Multiplication Property of Exponents

1) 4^6
2) 8^5
3) 7^6
4) 9^4
5) 2^7
6) 5^7
7) 4^7
8) $5x^2$
9) x^6
10) x^9
11) x^9
12) $30x^2$
13) $16x^6$
14) $7x^4$
15) $12x^6$
16) $5x^8$
17) $4x^9$
18) $6x^6$
19) $24x^8$
20) $20x^4y$
21) $7x^{10}y^5$
22) x^7y^7
23) $12x^8y^4$
24) $36x^6y^5$
25) $30x^{11}y^6$
26) $32x^4y^9$
27) $12x^4y^8$
28) $16x^7y^{10}$
29) $21x^3y^{13}$
30) $4x^7y^4$
31) $27x^6y^8$
32) $120x^9y^6$

Zero and Negative Exponents

1) 1
2) $\frac{1}{4}$
3) 0
4) 1
5) $\frac{1}{25}$
6) $\frac{1}{27}$
7) $\frac{1}{9}$
8) $\frac{1}{100}$
9) $\frac{1}{144}$
10) $\frac{1}{32}$
11) $\frac{1}{81}$
12) $\frac{1}{16}$
13) $\frac{1}{216}$
14) $\frac{1}{1,000}$
15) $\frac{1}{30}$
16) $\frac{1}{225}$
17) $\frac{1}{64}$
18) $\frac{1}{128}$
19) $\frac{1}{125}$
20) $\frac{1}{256}$
21) $\frac{1}{243}$
22) $\frac{1}{10,000}$
23) $\frac{1}{1,024}$
24) $\frac{1}{512}$
25) $\frac{1}{400}$
26) $\frac{1}{196}$
27) $\frac{1}{729}$
28) $\frac{1}{10,000}$
29) $\frac{1}{625}$
30) $\frac{1}{4,096}$
31) 64
32) 36
33) 49
34) $\frac{27}{8}$
35) 169
36) $\frac{144}{49}$
37) 216
38) $90,000$
39) $\frac{81}{4}$
40) $\frac{5}{7}$
41) 1
42) $1,024$

Division Property of Exponents

1) $\frac{1}{5}$
2) 8^2
3) 4^4
4) $\frac{1}{3^4}$
5) $\frac{1}{x^5}$
6) $\frac{1}{3^4}$
7) 9^2
8) 10
9) 1
10) $\frac{1}{2x^5}$
11) $\frac{3x^5}{4}$
12) $\frac{3}{2x}$
13) $\frac{7x^5}{y^9}$
14) $\frac{9y^3}{x^4}$
15) $\frac{2x^6}{9}$

16) $\frac{7y^6}{x}$ 18) $\frac{5}{x^2}$ 19) $\frac{1}{2x^5y^3}$ 21) $\frac{9}{4x^6}$

17) $\frac{2}{x^4y^{12}}$ 20) $\frac{1}{7}$

Powers of Products and Quotients

1) 4^6
2) 2^{12}
3) 2^8
4) 5^{36}
5) 19^{18}
6) 2^{28}
7) 5^6
8) 4^{16}
9) $64x^{10}$
10) $81x^8y^{16}$
11) $49x^{10}y^4$

12) $125x^{12}y^{12}$
13) $32x^{15}y^{15}$
14) $1,000x^9y^{12}$
15) $169y^8$
16) $25x^{20}$
17) $216x^{21}y^{18}$
18) $144x^{24}$
19) $256x^{20}$
20) $32x^{20}y^{15}$
21) $225x^{14}y^4$
22) $512x^9y^{15}$

23) $1,296x^4y^8$
24) $\frac{16}{x^8}$
25) x^9
26) $\frac{216y^3}{x^{12}}$
27) $\frac{1}{x^6y^{12}}$
28) x^6y^6
29) $\frac{25y^{16}}{x^4}$
30) $\frac{16}{y^{12}}$

Negative Exponents and Negative Bases

1) $-\frac{1}{9}$
2) $-\frac{1}{81}$
3) $-\frac{1}{32}$
4) $-\frac{1}{x^7}$
5) $\frac{11}{x}$
6) $-\frac{8}{x^3}$
7) $-\frac{12}{x^5}$
8) $-\frac{9}{x^8y^6}$
9) $\frac{32}{x^5y}$
10) $\frac{10}{a^9b^3}$

11) $-\frac{17x^4}{y^6}$
12) $-25x^5$
13) $-13xa^7$
14) 81
15) $\frac{16}{9}$
16) $-14a^6b^3$
17) $-7x^9$
18) $-\frac{b^5}{a^9}$
19) $-11x^5$
20) $-\frac{bc^6}{2}$
21) $12a^5b^4$

22) $-\frac{p^7}{4n^4}$
23) $-\frac{8ac^5}{3b^6}$
24) $\frac{c^4}{16a^4}$
25) $\frac{y^3z^3}{27x^3}$
26) $-\frac{8ac^3}{5b^7}$
27) $-x^5$
28) $49x^{10}$
29) x^{36}

Scientific Notation

1) 2.23×10^{-1}
2) 9×10^{-2}
3) 4.5×10^{0}
4) 9×10^{2}
5) 2×10^{3}
6) 6×10^{-3}
7) 3.3×10^{1}
8) 9.4×10^{3}
9) 1.47×10^{3}
10) 5.2×10^{4}

11) 8×10^{6}
12) 9×10^{-5}
13) 2.158×10^{6}
14) 3.9×10^{-3}
15) 7.5×10^{-5}
16) 4.3×10^{6}
17) 1.3×10^{5}
18) 4×10^{9}
19) 9×10^{-5}
20) 3.9×10^{-3}

21) 0.4
22) 0.0012
23) $270,000$
24) 0.0006
25) 0.0036
26) $550,000$
27) $32,000$
28) $3,880,000$
29) 0.000007
30) 0.00000042

Square Roots

1) 4
2) 5
3) 1
4) 8
5) 0
6) 14
7) 2
8) 16
9) 6

10) 17
11) 13
12) 12
13) 10
14) 40
15) 50
16) 18
17) 23
18) $2\sqrt{5}$

19) 25
20) $3\sqrt{2}$
21) $5\sqrt{2}$
22) 32
23) $4\sqrt{10}$
24) $4\sqrt{2}$
25) 10
26) 42
27) 6

28) 13
29) 30
30) 6
31) $2\sqrt{13}$
32) $3\sqrt{10}$
33) $2\sqrt{7}$
34) 80
35) 120
36) $4\sqrt{3}$

Simplifying radical expressions

1) $x\sqrt{13}$
2) $5x\sqrt{3}$
3) $3\sqrt[3]{a}$
4) $8x^2\sqrt{x}$
5) $6\sqrt{6a}$
6) $w\sqrt[3]{63}$
7) $8\sqrt{3x}$
8) $5\sqrt{5v}$

9) $4\sqrt[3]{2x^2}$
10) $10x^4\sqrt{x}$
11) $4x^2$
12) $5a\sqrt[3]{4a^2}$
13) $11\sqrt{2}$
14) $14p\sqrt{2p}$
15) $2m^3\sqrt{2}$
16) $3x.y\sqrt{22xy}$

17) $11x^2y^2\sqrt{xy}$
18) $4a^3b\sqrt{b}$
19) $3x^2y^3\sqrt{10xy}$
20) $4x^2\sqrt[3]{y^2}$
21) $40x^2$
22) $54x$
23) $2y^2\sqrt[3]{7x^2}$
24) $10xy^2\sqrt[3]{x^2y}$

25) $40\sqrt{2a}$

26) $5x^2\sqrt[4]{y}$

27) $2x^2y^2\mathrm{r}\sqrt{6yr}$

28) $30x^2y^2z^4\sqrt{y}$

29) $21x^3y^2\sqrt[3]{y}$

30) $45a^2bc^4\sqrt{ac}$

31) $5x^2y^4$

Multiplying radical expressions

1) 5

2) $5\sqrt{2}$

3) 6

4) $7\sqrt{47}$

5) -28

6) $15\sqrt{3}$

7) 48

8) $-5\sqrt{7}$

9) $11\sqrt{5}$

10) $504\sqrt{7}$

11) $15\sqrt{5} + 15$

12) $13x^2\sqrt{x}$

13) -18

14) $26x^4$

15) $7x^2\sqrt{2x}$

16) $-8x^3\sqrt{35}$

17) $-64x^4\sqrt{2}$

18) $-128\sqrt{2} - 128$

19) $40\sqrt{2x} - 8x$

20) $8x^3\sqrt{2} + 4\sqrt{x}$

21) $10\sqrt{5r} + 10\sqrt{r}$

22) $-84x^3\sqrt{2}$

23) $-12\sqrt{6}x$

24) $21v^2\sqrt{6}$

25) -14

26) $15\sqrt{5} - 27$

27) $40\sqrt{3} - 42$

28) $71 - 29\sqrt{5}$

29) $-3x - 5\sqrt{3x} - 4$

30) $14r - 15\sqrt{7r} + 25$

31) $7\sqrt{n} + 5\sqrt{7n} - \sqrt{7} - 5$

32) $-15 + 6\sqrt{3x} + 5\sqrt{3} - 6\sqrt{x}$

Simplifying radical expressions involving fractions

1) $\frac{\sqrt{15}}{3}$

2) $\frac{9\sqrt{10}}{45} = \frac{\sqrt{10}}{5}$

3) $\frac{\sqrt{20}}{10} = \frac{\sqrt{5}}{5}$

4) $\frac{13\sqrt{3}}{3}$

5) $\frac{12\sqrt{5mr}}{m^3}$

6) $\frac{11\sqrt{2k}}{k}$

7) $3x\sqrt{5}$

8) $\frac{\sqrt{2x}}{xy}$

9) $\frac{-1-\sqrt{5}}{4}$

10) $\frac{\sqrt{11a} - 8a\sqrt{11}}{11a}$

11) $\frac{a-\sqrt{ab}}{a-b}$

12) $\frac{\sqrt{30}+2\sqrt{5}-\sqrt{6}-2}{2}$

13) $12 + 8\sqrt{2} + 3\sqrt{7} + 2\sqrt{14}$

14) $-\frac{5(\sqrt{3}-1)}{6}$

15) $-14 - 7\sqrt{5}$

16) -1

17) $6 + \sqrt{35}$

18) $\frac{12 - 2\sqrt{10} - 3\sqrt{6} + \sqrt{15}}{13}$

19) $\frac{4\sqrt{11}+11+20\sqrt{3}+5\sqrt{33}}{5}$

20) $2\sqrt{5} + 3 + \sqrt{15} + 2\sqrt{3}$

21) $4a^3\sqrt{b}$

22) $2x\sqrt{21}$

Adding and subtracting radical expressions

1) $3\sqrt{2}$

2) $19\sqrt{2}$

3) 0

4) $15\sqrt{2}$

5) $10\sqrt{3}$

6) $-10\sqrt{2}$

7) -48

8) $18\sqrt{6}$

9) 0

10) $20\sqrt{5}$

11) $26\sqrt{3}$

12) 30

13) $3\sqrt{5}$

14) $2\sqrt{7}$

15) $13\sqrt{11}$

16) $-11\sqrt{3}$

17) 3

18) 0

19) $14\sqrt{7}$

20) $13\sqrt{3}$

21) $12\sqrt{3}$

22) $5\sqrt{30}$

23) $-16\sqrt{6}$

24) 0

25) $5y\sqrt{7}$

26) $5n\sqrt{5m}$

27) $-12\sqrt{3a}$

28) $-9\sqrt{15ab}$

29) $5x\sqrt{5y}$

30) $14\sqrt{7a}$

Chapter 5 :

Algebraic Expressions

Topics that you will practice in this chapter:

- ✓ Simplifying Variable Expressions
- ✓ Simplifying Polynomial Expressions
- ✓ Translate Phrases into an Algebraic Statement
- ✓ The Distributive Property
- ✓ Evaluating One Variable Expressions
- ✓ Evaluating Two Variables Expressions
- ✓ Combining like Terms

Mathematics is, as it were, a sensuous logic, and relates to philosophy as do the arts, music, and plastic art to poetry. — K. Shegel

Simplifying Variable Expressions

✎ Simplify each expression.

1) $3(x + 5) =$

2) $(-4)(7x - 5) =$

3) $11x + 5 - 6x =$

4) $-4 - 2x^2 - 6x^2 =$

5) $7 + 13x^2 + 3 =$

6) $3x^2 + 7x + 15x^2 =$

7) $3x^2 - 12x^2 + 4x =$

8) $4x^2 - 8x - 2x =$

9) $6x + 7(3 - 4x) =$

10) $8x + 4(15x - 3) =$

11) $6(-3x - 9) - 17 =$

12) $-11x^2 - (-5x) =$

13) $2x + 7 + 5 - 8x =$

14) $7 + 6x - 11 - 5x =$

15) $27x + 8 - 13 - 5x =$

16) $(-11)(-5x + 2) - 41x =$

17) $19x - 4(4 - 2x) =$

18) $16x + 3(3x + 6) + 10 =$

19) $5(-2x - 4) - 13x =$

20) $16x - 3x(x + 10) =$

21) $17x + 5x(2 - 4x) =$

22) $5x(-4x - 7) + 20x =$

23) $25x - 19 + 4x^2 =$

24) $6x(x - 11) + 25 =$

25) $4x - 5 + 15x + 3x^2 =$

26) $-7x^2 - 11x - 9x =$

27) $10x - 9x^2 - 3x^2 - 7 =$

28) $13 + 3x^2 - 9x^2 - 21x =$

29) $22x + 10x^2 - 15x + 17 =$

30) $4x^2 + 25x + 21x^2 =$

31) $29 - 12x^2 - 23x - 4x^2 =$

32) $22x - 19x - 9x^2 + 30 =$

Simplifying Polynomial Expressions

✎ **Simplify each polynomial.**

1) $(2x^3 + 8x^2) - (11x + 3x^2) = $ _____

2) $(2x^5 + 7x^3) - (5x^3 + 11x^2) = $ _____

3) $(41x^4 + 5x^2) - (4x^2 + 20x^4) = $ _____

4) $13x - 8x^2 + 4(4x^2 + 3x^3) = $ _____

5) $(4x^3 - 22) + 5(3x^2 - 6x^3) = $ _____

6) $(4x^3 - 3x) - 5(2x^3 + x^4) = $ _____

7) $5(5x - 2x^3) - 2(8x^3 + 5x^2) = $ _____

8) $(3x^2 - 10x) - (5x^3 + 14x^2) = $ _____

9) $5x^3 - (3x^4 + 5x) + 2x^2 = $ _____

10) $11x^4 - (3x^2 + 5x) + 7x = $ _____

11) $(6x^2 - 3x^4) - (10x^4 + 3x^2) = $ _____

12) $2x^2 - 7x^3 + 19x^4 - 22x^3 = $ _____

13) $10x^2 - x^4 + 4x^4 - 32x^3 = $ _____

14) $-5x^2 + 17x^3 - 8x^2 - 6x = $ _____

15) $x^4 - 11x^5 - 30x^4 + 5x^2 = $ _____

16) $21x^3 + 13x - 5x^2 - 11x^3 = $ _____

Translate Phrases into an Algebraic Statement

✍ **Write an algebraic expression for each phrase.**

1) 9 multiplied by x. _____

2) Subtract 11 from y. _____

3) 19 divided by x. _____

4) 38 decreased by y. _____

5) Add y to 40. _____

6) The square of 6. _____

7) x raised to the fifth power. _____

8) The sum of six and a number. _____

9) The difference between fifty–seven and y. _____

10) The quotient of nine and a number. _____

11) The quotient of the square of x and 25. _____

12) The difference between x and 6 is 19. _____

13) 10 times a reduced by the square of b. _____

14) Subtract the product of a and b from 41. _____

The Distributive Property

✎ Use the distributive property to simply each expression.

1) $4(1 + 2x) =$

2) $2(4 + 7x) =$

3) $3(4x - 4) =$

4) $(2x - 5)(-6) =$

5) $(-3)(x + 6) =$

6) $(4 + 3x)2 =$

7) $(-5)(8 - 3x) =$

8) $-(-5 - 7x) =$

9) $(-6x + 3)(-3) =$

10) $(-4)(x - 7) =$

11) $-(5 - 3x) =$

12) $3(9 + 4x) =$

13) $6(4 + 3x) =$

14) $(-5x + 3)2 =$

15) $(5 - 8x)(-3) =$

16) $(-12)(3x + 3) =$

17) $(5 - 3x)6 =$

18) $4(2 + 6x) =$

19) $8(7x - 3) =$

20) $(-2x + 3)4 =$

21) $(7 - 5x)(-9) =$

22) $(-10)(x - 8) =$

23) $(11 - 4x)3 =$

24) $(-6)(10x - 4) =$

25) $(3 - 9x)(-7) =$

26) $(-9)(x + 9) =$

27) $(-3 + 5x)(-7) =$

28) $(-5)(8 - 10x) =$

29) $12(4x - 8) =$

30) $(-10x + 13)(-3) =$

31) $(-8)(3x - 2) + 4(x + 5) =$

32) $(-8)(x + 4) - (6 + 5x) =$

Evaluating One Variable Expressions

✎ **Evaluate each expression using the value given.**

1) $8 - x, x = 5$

2) $x - 9, x = 5$

3) $5x + 4, x = 3$

4) $x - 13, x = -4$

5) $12 - x, x = 4$

6) $x + 2, x = 6$

7) $4x + 8, x = 3$

8) $x + (-7), x = -8$

9) $4x + 5, x = 2$

10) $3x + 9, x = -2$

11) $15 + 3x - 7, x = 2$

12) $17 - 3x, x = 3$

13) $8x - 9, x = 4$

14) $5x + 4, x = -3$

15) $10x + 5, x = 3$

16) $14 - 4x, x = -6$

17) $3(5x + 3), x = 9$

18) $4(-3x - 6), x = 3$

19) $7x - 2x + 12, x = 4$

20) $(5x + 6) \div 2, x = 8$

21) $(x + 18) \div 10, x = 12$

22) $5x - 12 + 3x, x = -3$

23) $(6 - 4x)(-3), x = -4$

24) $9x^2 + 3x - 6, x = 2$

25) $x^2 - 10x, x = -5$

26) $3x(7 - 2x), x = 2$

27) $12x + 6 - 2x^2, x = -4$

28) $(-3)(4x - 8 + 3x), x = 3$

29) $(-6) + \frac{x}{4} + 3x, x = 16$

30) $(-6) + \frac{x}{5}, x = 35$

31) $\left(-\frac{45}{x}\right) - 7 + 2x, x = 9$

32) $\left(-\frac{21}{x}\right) - 12 + 4x, x = 7$

Evaluating Two Variables Expressions

✎ **Evaluate each expression using the values given.**

1) $2x - 4y$,

  $x = 4, y = 1$

2) $3x + 5y$,

  $x = -2, y = 2$

3) $-7a + 4b$,

  $a = 2, b = 4$

4) $3x + 5 - y$,

  $x = 5, y = 6$

5) $3z + 12 - 2k$,

  $z = 5, k = 6$

6) $6(-x - 3y)$,

  $x = 5, y = -2$

7) $5a + 3b$,

  $a = 3, b = 4$

8) $7x \div 3y$,

  $x = 3, y = 7$

9) $2x + 15 + 5y$,

  $x = -3, y = 1$

10) $5a - (18 - b)$,

  $a = 2, b = 8$

11) $2z + 20 + 5k$,

  $z = -6, k = 5$

12) $xy + 10 + 4x$,

  $x = 3, y = 5$

13) $2x + 4y - 8 + 5$,

  $x = 5, y = 2$

14) $\left(-\frac{24}{x}\right) + 3 + 2y$,

  $x = 4, y = 6$

15) $(-3)(-3a - 3b)$,

  $a = 4, b = 5$

16) $12 + 4x - 7 - y$,

  $x = 3, y = 5$

17) $11x + 5 - 8y + 6$,

  $x = 5, y = 2$

18) $10 + 2(-4x - 5y)$,

  $x = 5, y = 4$

19) $5x + 13 + 6y$,

  $x = 5, y = 6$

20) $10a - (7a + 3b) - 11$,

  $a = 3, b = 8$

Combining like Terms

✎ **Simplify each expression.**

1) $11x + 3x + 6 =$

2) $8(2x - 6) =$

3) $18x - 7x + 11 =$

4) $(-4)(6x - 7) =$

5) $22x - 10x - 5 =$

6) $32x - 13 + 8x =$

7) $15 - (8x - 11) =$

8) $-24x + 17 - 11x =$

9) $12x - 8 - 6x + 9 =$

10) $21x + 5 - 36 + 12x =$

11) $28x + 3x - 11 =$

12) $(-3x + 4)5 =$

13) $2 + 4x + 9x - 8 =$

14) $6(2x - 5x) - 4 =$

15) $4(5x + 11) + 3x =$

16) $x - 14 - 11x =$

17) $5(10 + 9x) - 8x =$

18) $42x + 17 - 23x =$

19) $(-7x) + 19 + 20x =$

20) $(-7x) - 33 + 29x =$

21) $4(5x + 3) - 19x =$

22) $5(6 - 2x) - 15x =$

23) $-24x + (11 - 18x) =$

24) $(-9) - (6)(7x + 3) =$

25) $(-1)(8x - 10) - 21x =$

26) $-36x + 14 + 27x - 5x =$

27) $3(-13x + 6) - 17x =$

28) $-5x - 42 + 32x =$

29) $37x - 19x + 15 - 9x =$

30) $3(5x + 7x) - 31 =$

31) $14 - 6x - 15 - 9x =$

32) $-2(-5x - 7x) + 27x =$

Answers of Worksheets

Simplifying Variable Expressions

1) $3x + 15$

2) $-28x + 20$

3) $5x + 5$

4) $-8x^2 - 4$

5) $13x^2 + 10$

6) $18x^2 + 7x$

7) $-9x^2 + 4x$

8) $4x^2 - 10x$

9) $-22x + 21$

10) $68x - 12$

11) $-18x - 71$

12) $-11x^2 + 5x$

13) $-6x + 12$

14) $x - 4$

15) $22x - 5$

16) $14x - 22$

17) $27x - 16$

18) $25x + 28$

19) $-23x - 20$

20) $-3x^2 - 14x$

21) $-20x^2 + 27x$

22) $-20x^2 - 15x$

23) $4x^2 + 25x - 19$

24) $6x^2 - 66x + 25$

25) $3x^2 + 19x - 5$

26) $-7x^2 - 20x$

27) $-12x^2 + 10x - 7$

28) $-6x^2 - 21x + 13$

29) $10x^2 + 7x + 17$

30) $25x^2 + 25x$

31) $-16x^2 - 23x + 29$

32) $-9x^2 + 3x + 30$

Simplifying Polynomial Expressions

1) $2x^3 + 5x^2 - 11x$

2) $2x^5 + 2x^3 - 11x^2$

3) $21x^4 + x^2$

4) $12x^3 + 8x^2 + 13x$

5) $-26x^3 + 15x^2 - 22$

6) $-5x^4 - 6x^3 - 3x$

7) $-26x^3 - 10x^2 + 25x$

8) $-5x^3 - 11x^2 - 10x$

9) $-3x^4 + 5x^3 + 2x^2 - 5x$

10) $11x^4 - 3x^2 + 2x$

11) $-13x^4 + 3x^2$

12) $19x^4 - 29x^3 + 2x^2$

13) $3x^4 - 32x^3 + 10x^2$

14) $17x^3 - 13x^2 - 6x$

15) $-11x^5 - 29x^4 + 5x^2$

16) $10x^3 - 5x^2 + 13x$

Translate Phrases into an Algebraic Statement

1) $9x$

2) $y - 11$

3) $\frac{19}{x}$

4) $38 - y$

5) $y + 40$

6) 6^2

7) x^5

8) $6 + x$

9) $57 - y$

10) $\frac{9}{x}$

11) $\frac{x^2}{25}$

12) $x - 6 = 19$

13) $10a - b^2$

14) $41 - ab$

The Distributive Property

1) $8x + 4$

2) $14x + 8$

3) $12x - 12$

4) $-12x + 30$

5) $-3x - 18$

6) $6x + 8$

7) $15x - 40$

8) $7x + 5$

9) $18x - 9$

10) $-4x + 28$

11) $3x - 5$

12) $12x + 27$

13) $18x + 24$ 18) $24x + 8$ 23) $-12x + 33$ 28) $50x - 40$

14) $-10x + 6$ 19) $56x - 24$ 24) $-60x + 24$ 29) $48x - 96$

15) $24x - 15$ 20) $-8x + 12$ 25) $63x - 21$ 30) $30x - 39$

16) $-36x - 36$ 21) $45x - 63$ 26) $-9x - 81$ 31) $-20x + 36$

17) $-18x + 30$ 22) $-10x + 80$ 27) $-35x + 21$ 32) $-13x - 38$

Evaluating One Variables

1) 3   9) 13   17) 144   25) 75

2) −4   10) 3   18) −60   26) 18

3) 19   11) 14   19) 32   27) −74

4) −17   12) 8   20) 23   28) −39

5) 8   13) 23   21) 3   29) 46

6) 8   14) −11   22) −36   30) 1

7) 20   15) 35   23) −66   31) 6

8) − 15   16) 38   24) 36   32) 13

Evaluating Two Variables

1) 4   6) 6   11) 33   16) 12

2) 4   7) 27   12) 37   17) 50

3) 2   8) 1   13) 15   18) −70

4) 14   9) 14   14) 9   19) 74

5) 15   10) 0   15) 81   20) −26

Combining like Terms

1) $14x + 6$ 9) $6x + 1$ 17) $37x + 50$ 25) $-29x + 10$

2) $16x - 48$ 10) $33x - 31$ 18) $19x + 17$ 26) $-14x + 14$

3) $11x + 11$ 11) $31x - 11$ 19) $13x + 19$ 27) $-56x + 18$

4) $-24x + 28$ 12) $-15x + 20$ 20) $22x - 33$ 28) $27x - 42$

5) $12x - 5$ 13) $13x - 6$ 21) $x + 12$ 29) $9x + 15$

6) $40x - 13$ 14) $-18x - 4$ 22) $-25x + 30$ 30) $36x - 31$

7) $-8x + 26$ 15) $23x + 44$ 23) $-42x + 11$ 31) $-15x - 1$

8) $-35x + 17$ 16) $-10x - 14$ 24) $-42x - 27$ 32) $51x$

Chapter 6

Equations and Inequalities

Topics that you will practice in this chapter:

- ✓ One–Step Equations
- ✓ Multi–Step Equations
- ✓ Graphing Single–Variable Inequalities
- ✓ One–Step Inequalities
- ✓ Multi-Step Inequalities
- ✓ Systems of Equations
- ✓ Systems of Equations Word Problems

"Life is a math equation. In order to gain the most, you have to know how to convert negatives into positives." – Anonymous

One–Step Equations

✎ **Find the answer for each equation.**

1) $3x = 90, x =$ ____

2) $5x = 35, x =$ ____

3) $6x = 24, x =$ ____

4) $24x = 144, x =$ ____

5) $x + 15 = 20, x =$ ____

6) $x - 7 = 4, x =$ ____

7) $x - 9 = 2, x =$ ____

8) $x + 15 = 23, x =$ ____

9) $x - 4 = 13, x =$ ____

10) $12 = 16 + x, x =$ ____

11) $x - 10 = 2, x =$ ____

12) $5 - x = -11, x =$ ____

13) $28 = -6 + x, x =$ ____

14) $x - 20 = -35, x =$ ____

15) $x + 14 = -4, x =$ ____

16) $14 = 28 - x, x =$ ____

17) $7 + x = -7, x =$ ____

18) $x - 16 = 4, x =$ ____

19) $30 = x - 15, x =$ ____

20) $x - 5 = -18, x =$ ____

21) $x - 10 = 24, x =$ ____

22) $x - 20 = -25, x =$ ____

23) $x - 17 = 30, x =$ ____

24) $-70 = x - 28, x =$ ____

25) $x - 9 = 13, x =$ ____

26) $36 = 4x, x =$ ____

27) $x - 35 = 25, x =$ ____

28) $x - 25 = 10, x =$ ____

29) $70 - x = 16, x =$ ____

30) $x - 10 = 14, x =$ ____

31) $17 - x = -13, x =$ __

32) $x - 9 = -30, x =$ ____

Multi–Step Equations

✎ **Find the answer for each equation.**

1) $3x + 3 = 9$

2) $-x + 5 = 12$

3) $4x - 8 = 8$

4) $-(3 - x) = 5$

5) $4x - 8 = 16$

6) $12x - 15 = 9$

7) $2x - 18 = 2$

8) $4x + 8 = 16$

9) $24x + 27 = 75$

10) $-14(3 + x) = 14$

11) $-3(2 + x) = 6$

12) $12 = -(x - 7)$

13) $3(3 - x) = 30$

14) $-15 = -(3x + 6)$

15) $40(3 + x) = 40$

16) $5(x - 10) = 25$

17) $-18 = x + 8x$

18) $3x + 25 = -2x - 10$

19) $7(6 + 3x) = -63$

20) $18 - 3x = -4 - 5x$

21) $4 - 6x = 36 + 2x$

22) $15 + 15x = -5 + 5x$

23) $42 = (-6x) - 7 + 7$

24) $21 = 3x - 21 + 4x$

25) $-18 = -6x - 9 + 3x$

26) $5x - 15 = -29 + 6x$

27) $7x - 18 = 4x + 3$

28) $-7 - 4x = 5(4 - x)$

29) $x - 5 = -5(-3 - x)$

30) $13x - 68 = 15x - 102$

31) $-5x - 3 = -3(9 + 3x)$

32) $-2x - 15 = 6x + 17$

Graphing Single–Variable Inequalities

✎ **Draw a graph for each inequality.**

1) $x > -1$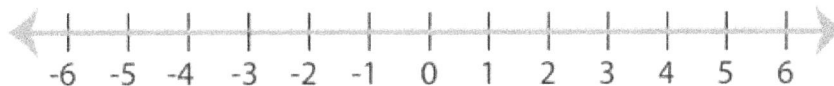

2) $x \leq 2$

3) $x \geq 0$

4) $x < -3$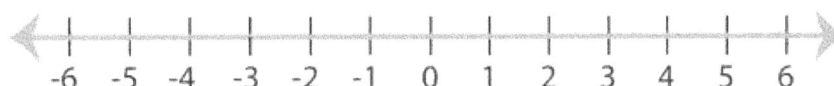

5) $x < \frac{1}{2}$

6) $x \leq -2$

7) $x \leq 3$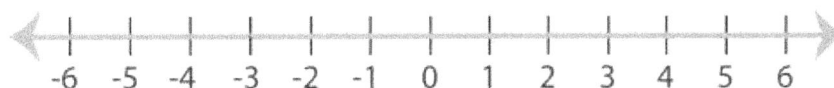

8) $x \geq -\frac{7}{2}$

One–Step Inequalities

✎ Find the answer for each inequality and graph it.

1) $x + 4 \geq 4$

2) $x - 5 \leq 2$

3) $5x > 35$

4) $9 + x \leq 11$

5) $x - 5 < -9$

6) $9x \geq 72$

7) $9x \leq 27$

8) $x + 19 > 16$

Multi-Step Inequalities

✎ **Calculate each inequality.**

1) $x - 3 \leq 7$

2) $8 - x \leq 8$

3) $3x - 9 \leq 9$

4) $4x - 4 \geq 8$

5) $x - 7 \geq 1$

6) $5x - 15 \leq 5$

7) $6x - 8 \leq 4$

8) $-11 + 6x \leq 12$

9) $4(x - 4) \leq 16$

10) $3x - 10 \leq 11$

11) $5x - 25 < 25$

12) $9x - 5 < 22$

13) $20 - 7x \geq -15$

14) $33 + 6x < 45$

15) $8 + 8x \geq 96$

16) $7 + 3x < 13$

17) $4x - 3 < 9$

18) $5(2 - 2x) \geq -30$

19) $-(7 + 6x) < 29$

20) $12 - 8x \geq -20$

21) $-4(x - 6) > 24$

22) $\dfrac{3x + 9}{6} \leq 10$

23) $\dfrac{4x - 10}{3} \leq 2$

24) $\dfrac{2x - 8}{3} > 2$

25) $8 + \dfrac{x}{6} < 9$

26) $\dfrac{9x}{7} - 4 < 5$

27) $\dfrac{15x + 45}{15} > 1$

28) $16 + \dfrac{x}{4} < 6$

Systems of Equations

✏️ **Calculate each system of equations.**

1) $-x + y = 2$ $x = \_\_\_$
   $-4x + 2y = 6$ $y = \_\_\_$

2) $-15x + 3y = -9$ $x = \_\_\_$
   $9x - 16y = 48$ $y = \_\_\_$

3) $y = -7$ $x = \_\_\_$
   $6x + 5y = 7$ $y = \_\_\_$

4) $3y = -9x + 15$ $x = \_\_\_$
   $5x - 4y = -3$ $y = \_\_\_$

5) $10x - 9y = -13$ $x = \_\_\_$
   $-5x + 3y = 11$ $y = \_\_\_$

6) $-12x - 16y = 20$ $x = \_\_\_$
   $6x - 12y = 30$ $y = \_\_\_$

7) $5x - 14y = -23$ $x = \_\_\_$
   $-18x + 21y = 24$ $y = \_\_\_$

8) $15x - 21y = -6$ $x = \_\_\_$
   $2x - 3y = -2$ $y = \_\_\_$

9) $-x + 3y = 3$ $x = \_\_\_$
   $-14x + 16y = -10$ $y = \_\_\_$

10) $x + 5y = 50$ $x = \_\_\_$
   $3x + 10y = 80$ $y = \_\_\_$

11) $6x - 7y = -8$ $x = \_\_\_$
   $-x - 4y = -9$ $y = \_\_\_$

12) $2x + 4y = -10$ $x = \_\_\_$
   $2x - 8y = 14$ $y = \_\_\_$

13) $4x + 3y = 12$ $x = \_\_\_$
   $5x - 3y = 15$ $y = \_\_\_$

14) $3x - 2y = 3$ $x = \_\_\_$
   $7x - 8y = 22$ $y = \_\_\_$

15) $3x + 2y = 5$ $x = \_\_\_$
   $-10x - 4y = -14$ $y = \_\_\_$

16) $10x + 7y = 1$ $x = \_\_\_$
   $-5x - 7y = 24$ $y = \_\_\_$

Systems of Equations Word Problems

✎ **Find the answer for each word problem.**

1) Tickets to a movie cost $4 for adults and $3 for students. A group of friends purchased 8 tickets for $31.00. How many adults ticket did they buy? ____

2) At a store, Eva bought two shirts and five hats for $77.00. Nicole bought three same shirts and four same hats for $84.00. What is the price of each shirt? _____

3) A farmhouse shelters 18 animals, some are pigs, and some are ducks. Altogether there are 66 legs. How many pigs are there? _____

4) A class of 214 students went on a field trip. They took 36 vehicles, some cars and some buses. If each car holds 5 students and each bus hold 22 students, how many buses did they take? _____

5) A theater is selling tickets for a performance. Mr. Smith purchased 5 senior tickets and 3 child tickets for $105 for his friends and family. Mr. Jackson purchased 3 senior tickets and 5 child tickets for $79. What is the price of a senior ticket? $_____

6) The difference of two numbers is 10. Their sum is 20. What is the bigger number? $_____

7) The sum of the digits of a certain two–digit number is 7. Reversing its digits increase the number by 9. What is the number? _____

8) The difference of two numbers is 11. Their sum is 25. What are the numbers? _____

9) The length of a rectangle is 5 meters greater than 2 times the width. The perimeter of rectangle is 28 meters. What is the length of the rectangle? _____

10) Jim has 25 nickels and dimes totaling $1.80. How many nickels does he have? _____

Answers of Worksheets

One–Step Equations

1) 30	9) 17	17) −14	25) 22
2) 7	10) −4	18) 20	26) 9
3) 4	11) 12	19) 45	27) 60
4) 6	12) 16	20) −13	28) 35
5) 5	13) 34	21) 34	29) 54
6) 11	14) −15	22) −5	30) 24
7) 11	15) −18	23) 47	31) 30
8) 8	16) 14	24) −42	32) −21

Multi–Step Equations

1) 2	9) 2	17) −2	25) 3
2) −7	10) −4	18) −7	26) 14
3) 4	11) −4	19) −5	27) 7
4) 8	12) −5	20) −11	28) 27
5) 6	13) −7	21) −4	29) −5
6) 2	14) 3	22) −2	30) 17
7) 10	15) −2	23) −7	31) −6
8) 2	16) 15	24) 6	32) −4

Graphing Single–Variable Inequalities

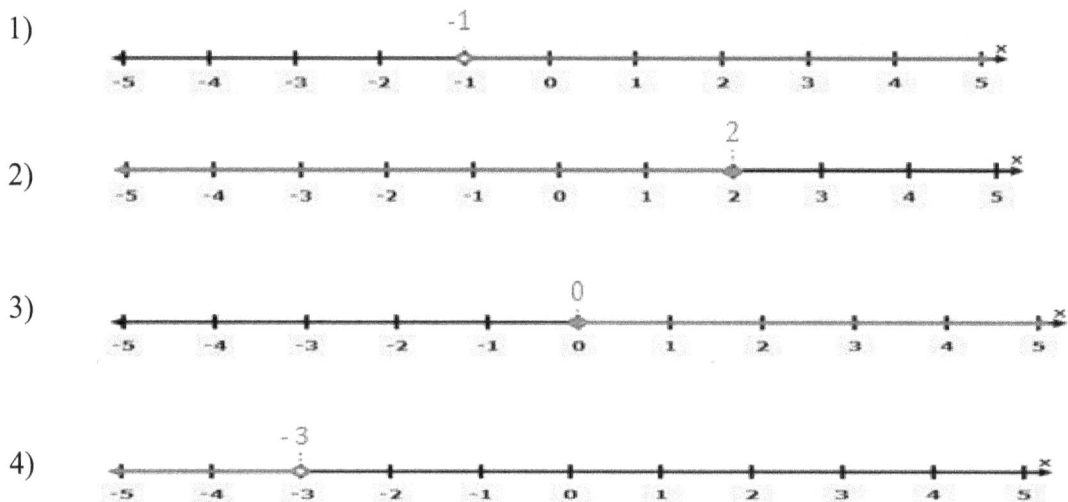

1)

2)

3)

4)

5)

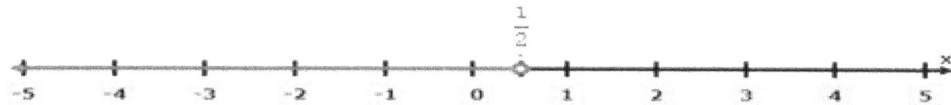

6)

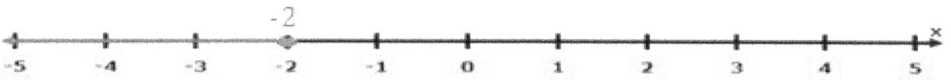

7)

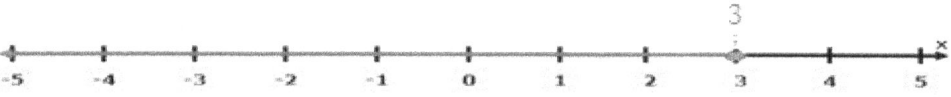

8)

One–Step Inequalities

1)

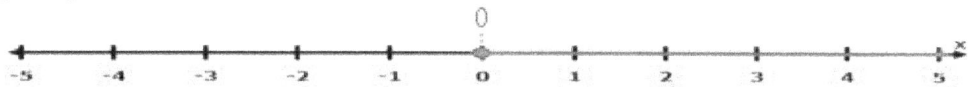

2)

3)

4)

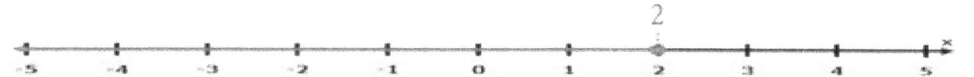

5)

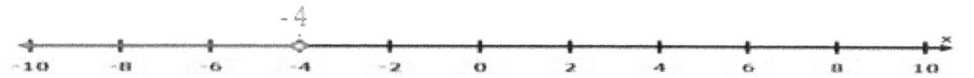

6)

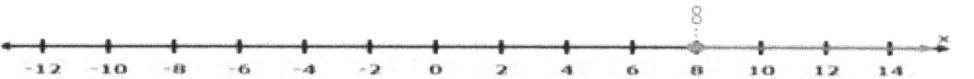

7)

8)

Multi-Step Inequalities

1) $x \leq 10$

2) $x \geq 0$

3) $x \leq 6$

4) $x \geq 3$

5) $x \geq 8$

6) $x \leq 4$

7) $x \leq 2$

8) $x \leq \frac{23}{6}$

9) $x \leq 8$

10) $x \leq 7$

11) $x < 10$

12) $x < 3$

13) $x \leq 5$

14) $x < 2$

15) $x \geq 11$

16) $x < 2$

17) $x < 3$

18) $x \leq 4$

19) $x > -6$

20) $x \leq 4$

21) $x < 0$

22) $x \leq 17$

23) $x \leq 4$

24) $x > 7$

25) $x < 6$ 26) $x < 7$ 27) $x > -2$ 28) $x < -40$

Systems of Equations

1) $x = -1, y = 1$ 7) $x = 1, y = 2$ 13) $x = 3, y = 0$

2) $x = 0, y = -3$ 8) $x = 8, y = 6$ 14) $x = -2, y = -\frac{9}{2}$

3) $x = 7$ 9) $x = 3, y = 2$ 15) $x = 1, y = 1$

4) $x = 1, y = 2$ 10) $x = -20, y = 14$ 16) $x = 5, y = -7$

5) $x = -4, y = -3$ 11) $x = 1, y = 2$

6) $x = 1, y = -2$ 12) $x = -1, y = -2$

Systems of Equations Word Problems

1) 7   5) $18   9) 11 meters

2) $16   6) 15   10) 14

3) 15   7) 34

4) 2   8) 18, 7

Chapter 7 :

Linear Functions

Topics that you will practice in this chapter:

✓ Finding Slope

✓ Graphing Lines Using Line Equation

✓ Writing Linear Equations

✓ Graphing Linear Inequalities

✓ Finding Midpoint

✓ Finding Distance of Two Points

"Nature is written in mathematical language." – Galileo Galilei

Finding Slope

🖎 Find the slope of each line.

1) $y = x + 8$

2) $y = -3x + 5$

3) $y = 2x + 12$

4) $y = -4x + 19$

5) $y = 11 + 6x$

6) $y = 7 - 5x$

7) $y = 8x + 19$

8) $y = -9x + 20$

9) $y = -7x + 4$

10) $y = 3x - 8$

11) $y = \frac{1}{3}x + 8$

12) $y = -\frac{4}{5}x + 9$

13) $-3x + 6y = 30$

14) $4x + 4y = 16$

15) $3y - x = 10$

16) $8y - x = 5$

🖎 Find the slope of the line through each pair of points.

17) $(2, 3), (7, 10)$

18) $(-3, 5), (2, 15)$

19) $(5, -3), (1, 9)$

20) $(-5, -5), (10, 25)$

21) $(22, 3), (7, 18)$

22) $(-16, 8), (-7, 26)$

23) $(25, 11), (29, 19)$

24) $(26, -19), (14, 17)$

25) $(22, -13), (20, -11)$

26) $(19, 7), (15, -3)$

27) $(5, 7), (11, 19)$

28) $(52, -62), (40, 70)$

Graphing Lines Using Line Equation

✍ **Sketch the graph of each line.**

1) $y = x - 2$

2) $y = -3x + 2$

3) $x + y = 0$

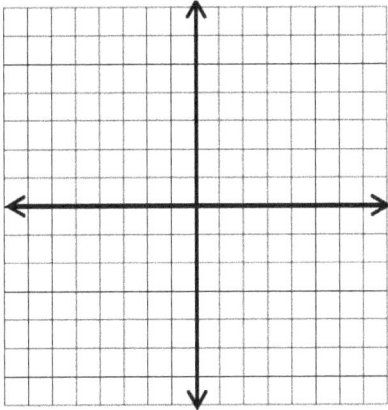 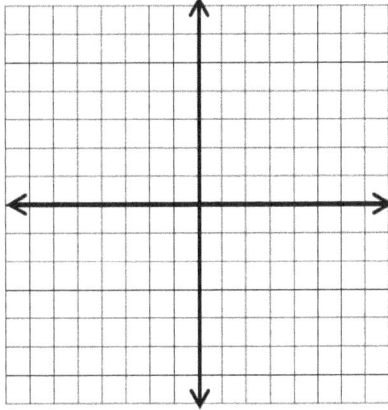 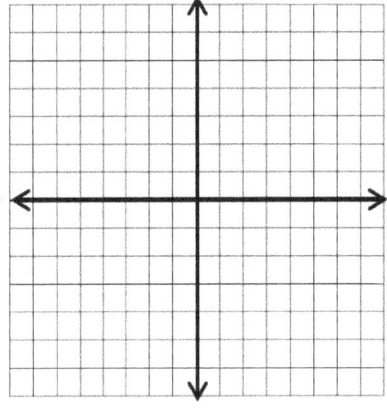

4) $x + y = -3$

5) $2x + 3y = -4$

6) $y - 3x + 6 = 0$

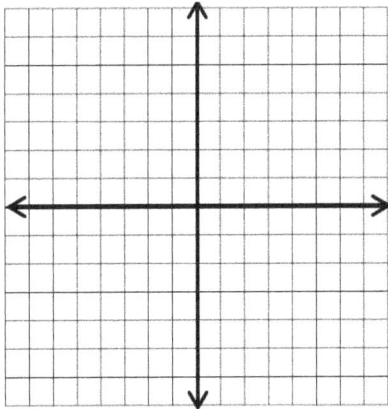 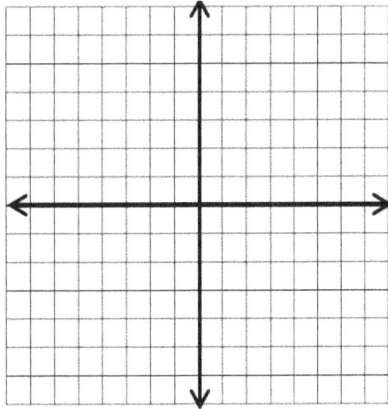 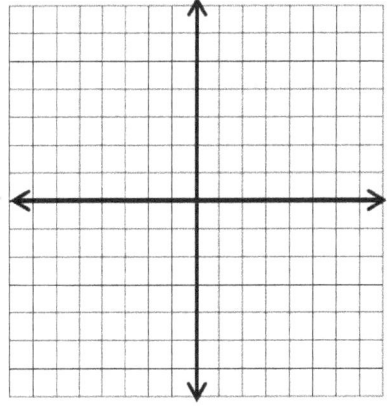

Writing Linear Equations

✎ **Write the equation of the line through the given points.**

1) Through: $(2, -5), (3, 9)$

2) Through: $(-6, 3), (3, 12)$

3) Through: $(10, 7), (5, 27)$

4) Through: $(15, 11), (3, -1)$

5) Through: $(24, 17), (12, -7)$

6) Through: $(8, 29), (4, -7)$

7) Through: $(20, -16), (12, 0)$

8) Through: $(-3, 10), (2, -5)$

9) Through: $(-6, 17), (4, -3)$

10) Through: $(-8, 22), (5, -4)$

11) Through: $(9, 27), (3, -3)$

12) Through: $(11, 32), (9, 4)$

13) Through: $(-3, 13), (-4, 0)$

14) Through: $(-5, 5), (5, 15)$

15) Through: $(18, -32), (11, 3)$

16) Through: $(-4, 25), (4, -15)$

✎ **Find the answer for each problem.**

17) What is the equation of a line with slope 6 and intercept 12?

_____

18) What is the equation of a line with slope -11 and intercept -4?

_____

19) What is the equation of a line with slope -3 and passes through point $(5, 2)$? _____

20) What is the equation of a line with slope -5 and passes through point $(-2, -1)$? _____

21) The slope of a line is -10 and it passes through point $(-3, 0)$. What is the equation of the line? _____

22) The slope of a line is 8 and it passes through point $(0, 7)$. What is the equation of the line? _____

Graphing Linear Inequalities

✎ **Sketch the graph of each linear inequality.**

1) $y > 4x - 5$ 2) $y < 2x + 4$ 3) $y \leq -5x - 2$

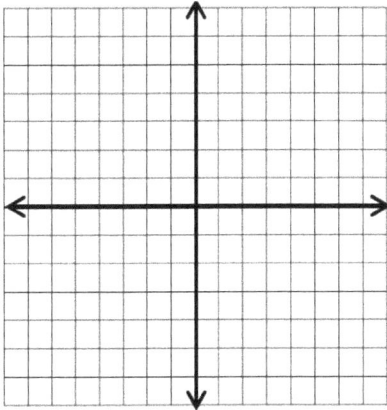 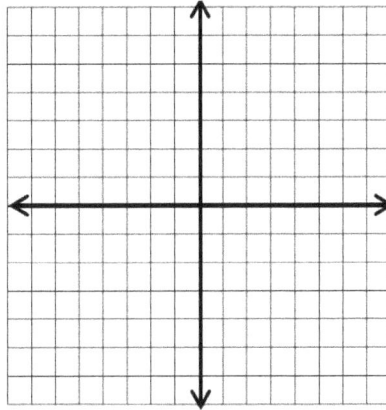 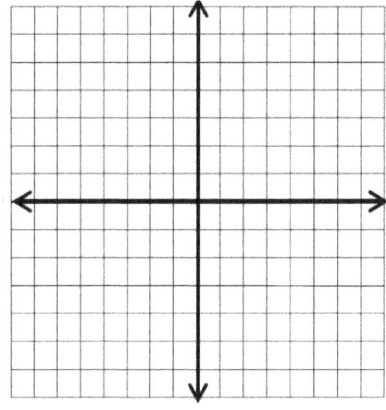

4) $4y \geq 12 + 4x$ 5) $-12y < 3x - 24$ 6) $5y \geq -15x + 10$

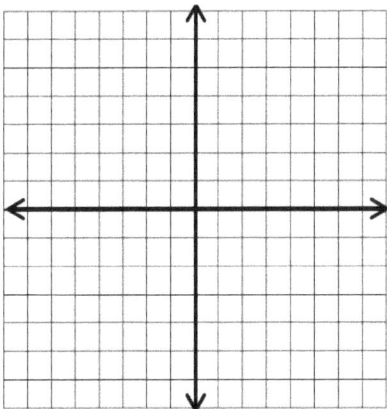 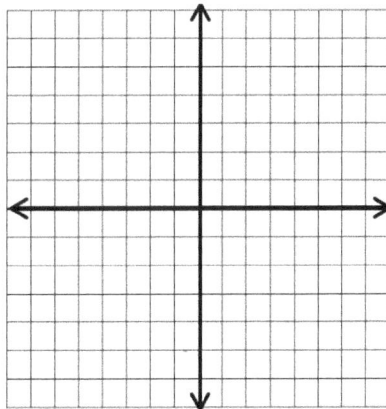 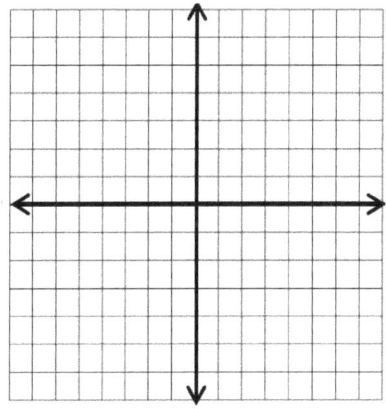

Finding Midpoint

✎ **Find the midpoint of the line segment with the given endpoints.**

1) $(-4, -3), (2, 3)$

2) $(9, 0), (-1, 8)$

3) $(9, -6), (3, 14)$

4) $(-10, -6), (0, 8)$

5) $(2, -5), (14, -15)$

6) $(-10, -3), (4, -13)$

7) $(8, 7), (-8, 13)$

8) $(-3, 6), (-9, 2)$

9) $(-4, 5), (16, -9)$

10) $(7, 14), (9, -2)$

11) $(-8, 6), (6, 6)$

12) $(10, 5), (-2, -3)$

13) $(-5, 12), (-3, 3)$

14) $(12, 7), (8, -2)$

15) $(10, 2), (-6, 14)$

16) $(-1, -2), (-7, 10)$

17) $(7, -7), (13, -13)$

18) $(-3, -8), (11, -4)$

19) $(5, -11), (-8, 9)$

20) $(14, -4), (16, 14)$

21) $(0, -5), (8, -1)$

22) $(3, 0), (-21, 18)$

23) $(17, -3), (-7, -5)$

24) $(26, -12), (6, 24)$

✎ **Find the answer for each problem.**

25) One endpoint of a line segment is $(-3, 7)$ and the midpoint of the line segment is $(-6, 9)$. What is the other endpoint? _____

26) One endpoint of a line segment is $(-3, 7)$ and the midpoint of the line segment is $(1, 5)$. What is the other endpoint? _____

27) One endpoint of a line segment is $(-10, -16)$ and the midpoint of the line segment is $(2, 9)$. What is the other endpoint? _____

Finding Distance of Two Points

✎ **Find the distance between each pair of points.**

1) $(6, 3), (-3, -9)$

2) $(5, 2), (-10, -6)$

3) $(8, 5), (8, 3)$

4) $(-8, -2), (2, 22)$

5) $(6, -7), (-3, -7)$

6) $(12, 0), (-9, -20)$

7) $(3, 20), (3, -5)$

8) $(10, 17), (5, 5)$

9) $(7, -2), (-4, -2)$

10) $(13, 4), (5, -2)$

11) $(11, 13), (5, 5)$

12) $(1, 4), (-23, -3)$

13) $(9, 8), (5, -4)$

14) $(-11, -4), (5, 8)$

15) $(-2, -6), (-2, -12)$

16) $(-1, -4), (23, 3)$

17) $(19, 3), (7, -6)$

18) $(-5, -2), (3, 4)$

19) $(2, 6), (2, -12)$

20) $(-4, -2), (8, -2)$

✎ **Find the answer for each problem.**

21) Triangle ABC is a right triangle on the coordinate system and its vertices are $(-2, 5)$, $(-2, 1)$, and $(1, 1)$. What is the area of triangle ABC? _____

22) Three vertices of a triangle on a coordinate system are $(3, -6)$, $(-5, -12)$, and $(3, -18)$. What is the perimeter of the triangle? _____

23) Four vertices of a rectangle on a coordinate system are $(-2, 2)$, $(-2, 6)$, $(4, 2)$, and $(4, 6)$. What is its perimeter? _____

SSAT Upper-Level Subject Test Mathematics

Answers of Worksheets

Finding Slope

1) 1
2) -3
3) 2
4) -4
5) 6
6) -5
7) 8
8) -9

9) -7
10) 3
11) $\frac{1}{3}$
12) $-\frac{4}{5}$
13) $\frac{1}{2}$
14) -1
15) $\frac{1}{3}$

16) $\frac{1}{8}$
17) $\frac{7}{5}$
18) 2
19) -3
20) 2
21) -1
22) 2

23) 2
24) -3
25) -1
26) $\frac{5}{2}$
27) 2
28) -11

Graphing Lines Using Line Equation

1) $y = x - 2$

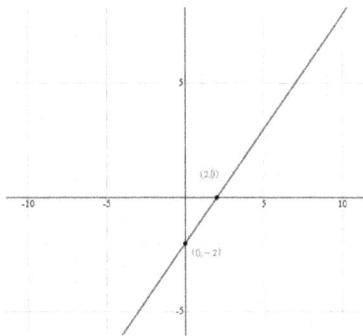

2) $y = -3x + 2$

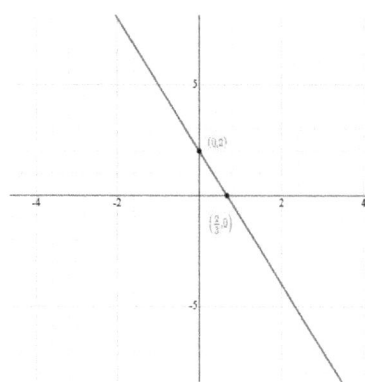

3) $x + y = 0$

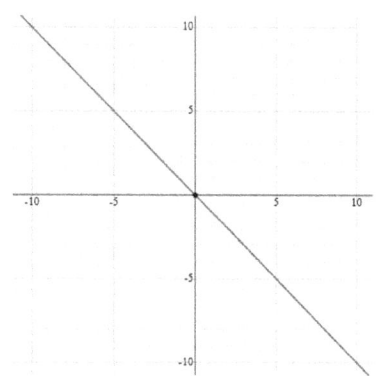

4) $x + y = -3$

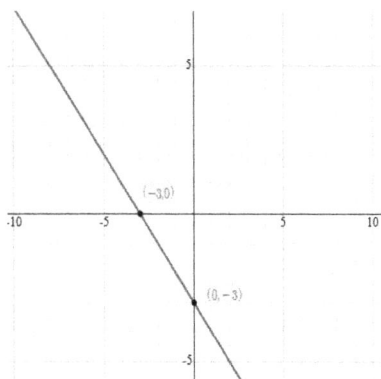

5) $2x + 3y = -4$

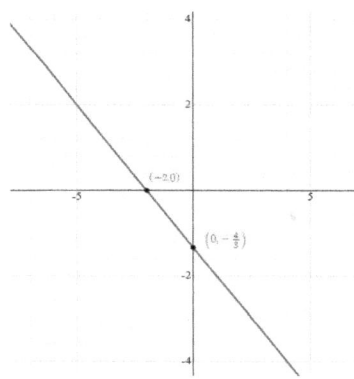

6) $y - 3x + 6 = 0$

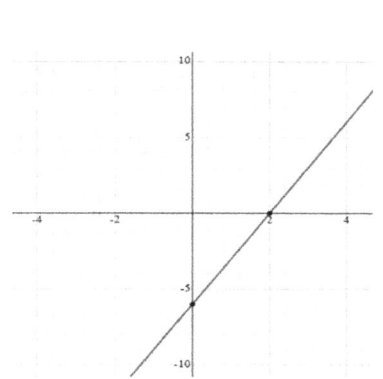

Writing Linear Equations

1) $y = 14x - 33$

2) $y = x + 9$

3) $y = -4x + 47$

4) $y = x - 4$

5) $y = 2x - 31$

6) $y = 9x - 43$

7) $y = -2x + 24$

8) $y = -3x + 1$

9) $y = -2x + 5$

10) $y = -2x + 6$

11) $y = 5x - 18$

12) $y = 14x - 122$

13) $y = 13x + 52$

14) $y = x + 10$

15) $y = -5x + 58$

16) $y = -5x + 5$

17) $y = 6x + 12$

18) $y = -11x - 4$

19) $y = -3x + 17$

20) $y = -5x - 11$

21) $y = -10x - 30$

22) $y = 8x + 7$

Graphing Linear Inequalities

1) $y > 4x - 5$

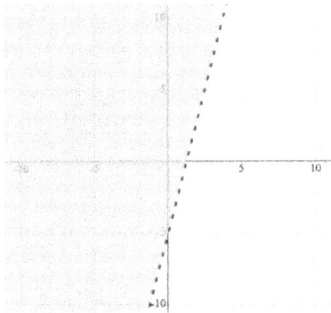

2) $y < 2x + 4$

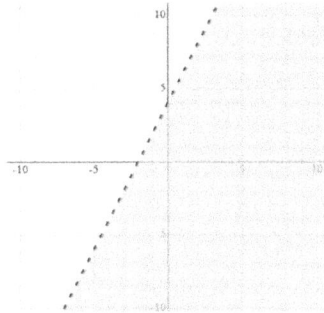

3) $y \leq -5x - 2$

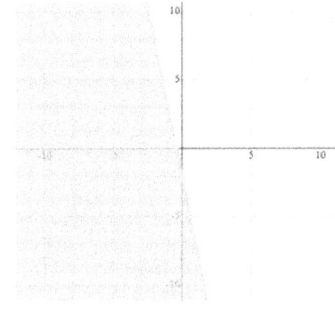

4) $4y \geq 12 + 4x$

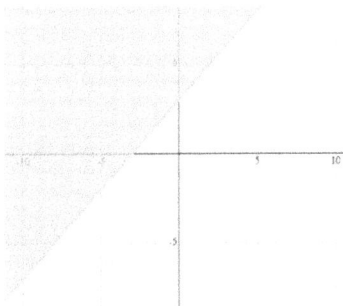

5) $-12y < 3x - 24$

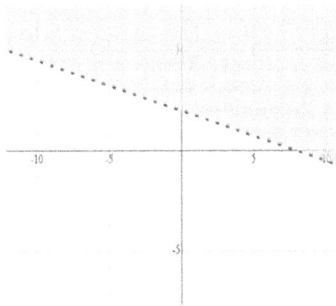

6) $5y \geq -15x + 10$

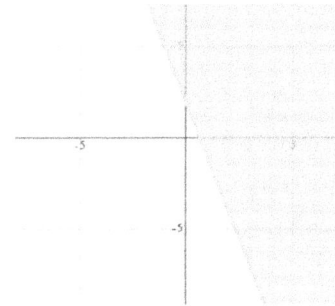

Finding Midpoint

1) $(-1, 0)$

2) $(4, 4)$

3) $(6, 4)$

4) $(-5, 1)$

5) $(8, -10)$

6) $(-3, -8)$

7) $(0, 10)$

8) $(-6, 4)$

9) $(6, -2)$

10) $(8, 6)$

11) $(-1, 6)$

12) $(4, 1)$

13) $(-4, 7.5)$

14) $(10, 2.5)$

15) $(2, 8)$

16) $(-4, 4)$

17) $(10, -10)$

18) $(4, -6)$

19) $(-1.5, -1)$

20) $(15, 5)$

21) $(4, -3)$

22) $(-9, 9)$

23) $(5, -4)$

24) $(16, 6)$

25) $(-9, 11)$

26) $(5, 3)$

27) $(14, 34)$

Finding Distance of Two Points

1) 15

2) 17

3) 2

4) 26

5) 9

6) 29

7) 25

8) 13

9) 11

10) 10

11) 10

12) 25

13) $4\sqrt{10}$

14) 20

15) 6

16) 25

17) 15

18) 10

19) 18

20) 12

21) 6 *square units*

22) 32 *units*

23) 20 *units*

Chapter 8 :

Polynomials

Topics that you will practice in this chapter:

✓ Writing Polynomials in Standard Form

✓ Simplifying Polynomials

✓ Adding and Subtracting Polynomials

✓ Multiplying Monomials

✓ Multiplying and Dividing Monomials

✓ Multiplying a Polynomial and a Monomial

✓ Multiplying Binomials

✓ Factoring Trinomials

✓ Operations with Polynomials

Mathematics is the supreme judge; from its decisions there is no appeal. – Tobias Dantzig

Writing Polynomials in Standard Form

✎ **Write each polynomial in standard form.**

1) $11x - 7x =$

2) $-5 + 19x - 19x =$

3) $6x^5 - 12x^3 =$

4) $12 + 17x^4 - 12 =$

5) $5x^2 + 4x - 9x^3 =$

6) $-3x^2 + 12x^5 =$

7) $5x + 8x^3 - 2x^8 =$

8) $-7x^3 + 4x - 9x^6 =$

9) $3x^2 + 22 - 6x =$

10) $3 - 4x + 9x^4 =$

11) $13x^2 + 28x - 8x^3 =$

12) $16 + 4x^2 - 2x^3 =$

13) $19x^2 - 9x + 9x^4 =$

14) $3x^4 - 7x^2 - 2x^3 =$

15) $-51 + 3x^2 - 8x^4 =$

16) $7x^2 - 8x^6 + 4x^4 - 15 =$

17) $6x^4 - 4x^5 + 16 - 3x^3 =$

18) $-2x^6 + 4x - 7x^2 - 5x =$

19) $11x^7 + 8x^5 - 5x^7 - 3x^2 =$

20) $2x^2 - 12x^5 + 8x^2 + 3x^6 =$

21) $4x^5 - 11x^7 - 6x^3 + 16x^5 =$

22) $6x^3 + 3x^5 + 34x^4 - 8x^5 =$

23) $3x(4x + 5 - 2x^2) =$

24) $12x(x^6 + 4x^3) =$

25) $5x(3x^2 + 6x + 4) =$

26) $7x(4 - 2x + 6x^5) =$

27) $3x(4x^4 - 4x^3 + 2) =$

28) $4x(2x^5 + 6x^2 - 3) =$

29) $5x(3x^4 + 4x^3 + 2x) =$

30) $2x(3x - 2x^3 + 4x^6) =$

Simplifying Polynomials

✎ **Simplify each expression.**

1) $3(4x - 20) =$

2) $5x(3x - 4) =$

3) $6x(5x - 7) =$

4) $3x(7x + 5) =$

5) $5x(4x - 3) =$

6) $6x(8x + 2) =$

7) $(3x - 2)(x - 4) =$

8) $(x - 5)(2x + 6) =$

9) $(x - 3)(x - 7) =$

10) $(3x + 4)(3x - 4) =$

11) $(5x - 4)(5x - 2) =$

12) $6x^2 + 6x^2 - 8x^4 =$

13) $3x - 2x^2 + 5x^3 + 7 =$

14) $7x + 4x^2 - 10x^3 =$

15) $12x^2 + 5x^5 - 6x^3 =$

16) $-5x^2 + 4x^6 + 6x^8 =$

17) $-12x^3 + 10x^5 - 4x^6 + 4x =$

18) $11 - 7x^2 + 4x^2 - 16x^3 + 11 =$

19) $2x^2 - 9x + 4x^3 + 15x - 10x =$

20) $13 - 7x^5 + 6x^5 - 4x^2 + 5 =$

21) $-5x^8 + x^6 - 14x^3 + 5x^8 =$

22) $(7x^4 - 4) + (7x^4 - 2x^4) =$

23) $3(3x^4 - 4x^3 - 6x^4) =$

24) $-5(x^9 + 8) - 5(10 - x^9) =$

25) $8x^3 - 9x^4 - 2x + 19 - 8x^3 =$

26) $11 - 8x^3 + 6x^3 - 7x^5 + 6 =$

27) $(5x^3 - 4x) - (6x - 2 - 6x^3) =$

28) $4x^2 - 5x^4 - x(3x^3 + 2x) =$

29) $6x + 6x^5 - 10 - 4(x^5 - 3) =$

30) $4 - 3x^4 + (6x^5 - 2x^4 + 5x^5) =$

31) $-(x^5 + 4) - 8(3 + x^5) =$

32) $(4x^3 - 3x) - (3x - 5x^3) =$

Adding and Subtracting Polynomials

✎ **Add or subtract expressions.**

1) $(-2x^2 - 3) + (3x^2 + 4) =$

2) $(4x^3 + 6) - (7 - 2x^3) =$

3) $(4x^5 + 5x^2) - (2x^5 + 15) =$

4) $(6x^3 - 2x^2) + (5x^2 - 4x) =$

5) $(10x^4 + 28x) - (34x^4 + 6) =$

6) $(7x^2 - 3) + (7x^2 + 3) =$

7) $(9x^2 + 4) - (10 - 5x^2) =$

8) $(6x^2 + x^5) - (x^5 + 4) =$

9) $(4x^3 - x) + (3x - 7x^3) =$

10) $(11x + 10) - (8x + 10) =$

11) $(15x^3 - 3x) - (3x - 4x^3) =$

12) $(4x - x^5) - (6x^5 + 8x) =$

13) $(2x^2 - 7x^7) - (4x^7 - 6x) =$

14) $(3x^2 - 5) + (8x^2 + 4x^5) =$

15) $(9x^4 + 5x^5) - (x^5 - 9x^4) =$

16) $(-4x^3 - 2x) + (9x - 5x^3) =$

17) $(4x - 3x^2) - (148x^2 + x) =$

18) $(5x - 8x^4) - (3x^4 - 4x^2) =$

19) $(8x^4 - 4) + (2x^4 - 3x^2) =$

20) $(5x^6 + 7x^3) - (x^3 - 5x^6) =$

21) $(-2x^2 + 20x^5 + 5x^4) + (12x^4 + 8x^5 + 24x^2) =$

22) $(7x^4 - 9x^7 - 6x) - (-3x^4 - 9x^7 + 6x) =$

23) $(14x + 12x^4 - 18x^6) + (20x^4 + 18x^6 - 10x) =$

24) $(5x^8 - 6x^6 - 4x) - (5x^3 + 9x^6 - 7x) =$

25) $(11x^2 - 6x^4 - 3x) - (-4x^2 - 12x^4 + 9x) =$

26) $(-5x^9 + 14x^3 + 3x^7) + (10x^7 + 26x^3 + 3x^9) =$

Multiplying Monomials

✎ **Simplify each expression.**

1) $6u^8 \times (-u^2) =$

2) $(-5p^8) \times (-2p^3) =$

3) $4xy^3z^5 \times 3z^4 =$

4) $3u^5t \times 8ut^4 =$

5) $(-5a^2) \times (-7a^3b^6) =$

6) $-3a^4b^3 \times 6a^2b =$

7) $13xy^5 \times x^4y^4 =$

8) $6p^4q^3 \times (-8pq^6) =$

9) $8s^4t^3 \times 4st^3 =$

10) $(-6x^4y^3) \times 6x^2y =$

11) $3xy^7z \times 12z^3 =$

12) $24xy \times x^2y =$

13) $13pq^4 \times (-3p^2q) =$

14) $13s^3t^4 \times st^4 =$

15) $11p^5 \times (-6p^3) =$

16) $(-8p^3q^5r) \times 3pq^4r^6 =$

17) $(-4a^4) \times (-7a^3b) =$

18) $6u^6v^2 \times (-5u^3v^4) =$

19) $9u^5 \times (-3u) =$

20) $-6xy^5 \times 4x^2y =$

21) $13y^5z^3 \times (-y^3z) =$

22) $8a^4bc^3 \times 2abc^3 =$

23) $(-7p^5q^6) \times (-5p^4q^2) =$

24) $4u^5v^3 \times (-4u^7v^3) =$

25) $17y^4z^5 \times (-y^6z) =$

26) $(-5pq^3r^2) \times 8p^2q^4r =$

27) $3ab^5c^6 \times 5a^4bc^2 =$

28) $6x^3yz^2 \times 3x^2y^7z^3 =$

Multiplying and Dividing Monomials

✎ Simplify each expression.

1) $(5x^5)(2x^2) =$

2) $(4x^4)(6x^2) =$

3) $(3x^4)(7x^4) =$

4) $(5x^6)(4x^2) =$

5) $(12x^4)(3x^6) =$

6) $(4yx^8)(8y^4x^3) =$

7) $(14x^4y)(x^3y^5) =$

8) $(-5x^3y^4)(2x^3y^5) =$

9) $(-6x^4y^2)(-3x^3y^5) =$

10) $(5x^3y)(-5x^2y^3) =$

11) $(6x^4y^3)(4x^3y^4) =$

12) $(4x^3y^2)(5x^2y^4) =$

13) $(12x^3y^6)(4x^4y^{10}) =$

14) $(15x^3y^5)(3x^4y^6) =$

15) $(7x^2y^7)(8x^6y^7) =$

16) $(-3x^3y^8)(7x^9y^4) =$

17) $\dfrac{5x^6y^6}{xy^4} =$

18) $\dfrac{19x^7y^5}{19x^6y} =$

19) $\dfrac{56x^4y^4}{8xy} =$

20) $\dfrac{81x^5y^6}{9x^4y^5} =$

21) $\dfrac{36x^7y^6}{9x^2y^3} =$

22) $\dfrac{48x^9y^7}{4x^4y^6} =$

23) $\dfrac{88x^{18}y^{12}}{11x^8y^9} =$

24) $\dfrac{30x^7y^6}{6x^8y^3} =$

25) $\dfrac{150x^7y^6}{30x^4y^6} =$

26) $\dfrac{-42x^{18}y^{14}}{6x^4y^9} =$

27) $\dfrac{-36x^7y^8}{9x^5y^8} =$

Multiplying a Polynomial and a Monomial

✏ **Find each product.**

1) $x(2x + 4) =$

2) $6(4 - 2x) =$

3) $5x(4x + 2) =$

4) $x(-4x + 5) =$

5) $8x(2x - 2) =$

6) $6(2x - 4y) =$

7) $7x(5x - 5) =$

8) $3x(12x + 2y) =$

9) $4x(x + 6y) =$

10) $11x(3x + 4y) =$

11) $7x(3x + 2) =$

12) $10x(4x - 10y) =$

13) $9x(3x - 2y) =$

14) $7x(x - 4y + 6) =$

15) $8x(2x^2 + 5y^2) =$

16) $12x(2x + 3y) =$

17) $4(2x^4 - 4y^4) =$

18) $4x(-3x^2y + 4y) =$

19) $-4(5x^3 - 2xy + 4) =$

20) $4(x^2 - 5xy - 6) =$

21) $8x(2x^3 - 5xy + 2x) =$

22) $-6x(-2x^3 - 6x + 2xy) =$

23) $3(2x^2 + xy - 9y^2) =$

24) $4x(5x^3 - 3x + 7) =$

25) $6(3x^{22} - 2x - 5) =$

26) $x^2(-2x^3 + 4x + 3) =$

27) $x^2(4x^3 + 10 - 2x) =$

28) $4x^4(3x^3 - 2x + 5) =$

29) $2x^2(4x^4 - 5xy + 7y^3) =$

30) $5x^2(5x^4 - 3x + 9) =$

31) $7x^2(6x^2 + 3x - 6) =$

32) $4x(x^3 - 4xy + 2y^2) =$

Multiplying Binomials

✎ **Find each product.**

1) $(x + 3)(x + 6) =$

2) $(x - 4)(x + 3) =$

3) $(x - 3)(x - 8) =$

4) $(x + 8)(x + 9) =$

5) $(x - 2)(x - 12) =$

6) $(x + 5)(x + 5) =$

7) $(x - 6)(x + 7) =$

8) $(x - 8)(x - 3) =$

9) $(x + 7)(x + 12) =$

10) $(x - 4)(x + 8) =$

11) $(x + 8)(x + 8) =$

12) $(x + 2)(x + 7) =$

13) $(x - 6)(x + 6) =$

14) $(x - 5)(x + 5) =$

15) $(x + 11)(x + 11) =$

16) $(x + 6)(x + 9) =$

17) $(x - 2)(x + 2) =$

18) $(x - 4)(x + 7) =$

19) $(3x + 5)(x + 6) =$

20) $(5x - 6)(4x + 8) =$

21) $(x - 7)(3x + 7) =$

22) $(x - 9)(x - 4) =$

23) $(x - 12)(x + 2) =$

24) $(2x - 4)(5x + 4) =$

25) $(3x - 8)(x + 8) =$

26) $(7x - 2)(6x + 3) =$

27) $(4x + 5)(3x + 5) =$

28) $(7x - 4)(9x + 4) =$

29) $(x + 2)(2x - 8) =$

30) $(5x - 4)(5x + 4) =$

31) $(3x + 2)(3x - 7) =$

32) $(x^2 + 8)(x^2 - 8) =$

Factoring Trinomials

✎ **Factor each trinomial.**

1) $x^2 + 8x + 12 =$

2) $x^2 - 6x + 5 =$

3) $x^2 + 15x + 36 =$

4) $x^2 - 12x + 35 =$

5) $x^2 - 11x + 18 =$

6) $x^2 - 9x + 18 =$

7) $x^2 + 18x + 72 =$

8) $x^2 - x - 72 =$

9) $x^2 + 4x - 21 =$

10) $x^2 - 13x + 22 =$

11) $x^2 + 2x - 24 =$

12) $x^2 - 3x - 40 =$

13) $x^2 - 3x - 70 =$

14) $x^2 + 26x + 169 =$

15) $4x^2 - 7x - 15 =$

16) $x^2 - 14x + 33 =$

17) $10x^2 + 5x - 15 =$

18) $6x^2 - 4x - 42 =$

19) $x^2 + 12x + 36 =$

20) $5x^2 + 17x - 12 =$

✎ **Calculate each problem.**

21) The area of a rectangle is $x^2 - x - 56$. If the width of rectangle is $x + 7$, what is its length? _____

22) The area of a parallelogram is $4x^2 + 17x - 15$ and its height is $x + 5$. What is the base of the parallelogram? _____

23) The area of a rectangle is $6x^2 - 22x + 12$. If the width of the rectangle is $3x - 2$, what is its length? _____

Operations with Polynomials

✏ **Find each product.**

1) $4(5x + 3) =$ _____

2) $8(2x + 6) =$ _____

3) $2(5x - 2) =$ _____

4) $-4(7x - 3) =$ _____

5) $3x^2(9x + 1) =$ _____

6) $4x^6(7x - 9) =$ _____

7) $3x^4(-7x + 3) =$ _____

8) $-8x^4(5x - 8) =$ _____

9) $7(x^2 + 5x - 3) =$ _____

10) $9(5x^2 - 7x + 5) =$ _____

11) $3(3x^2 + 3x + 2) =$ _____

12) $5x(3x^2 + 5x + 8) =$ _____

13) $(5x + 7)(3x - 3) =$ _____

14) $(9x + 3)(3x - 5) =$ _____

15) $(6x + 3)(4x - 2) =$ _____

16) $(7x - 2)(3x + 5) =$ _____

✏ **Calculate each problem.**

17) The measures of two sides of a triangle are $(2x + 5y)$ and $(6x - 3y)$. If the perimeter of the triangle is $(13x + 4y)$, what is the measure of the third side? _____

18) The height of a triangle is $(8x + 5)$ and its base is $(4x - 3)$. What is the area of the triangle? _____

19) One side of a square is $(6x + 2)$. What is the area of the square? _____

20) The length of a rectangle is $(5x - 8y)$ and its width is $(15x + 8y)$. What is the perimeter of the rectangle? _____

21) The side of a cube measures $(x + 2)$. What is the volume of the cube? _____

22) If the perimeter of a rectangle is $(28x + 6y)$ and its width is $(5x + 2y)$, what is the length of the rectangle? _____

Answers of Worksheets

Writing Polynomials in Standard Form

1) $4x$

2) -5

3) $6x^5 - 12x^3$

4) $14x^4$

5) $-9x^3 + 5x^2 + 4x$

6) $12x^5 - 3x^2$

7) $-2x^8 + 8x^3 + 5x$

8) $-9x^6 - 7x^3 + 4x$

9) $3x^2 - 6x + 22$

10) $9x^4 - 4x + 3$

11) $-8x^3 + 13x^2 + 28x$

12) $-2x^3 + 4x^2 + 16$

13) $9x^4 + 19x^2 - 9x$

14) $3x^4 - 2x^3 - 7x^2$

15) $-8x^4 + 3x^2 - 51$

16) $-8x^6 + 4x^4 + 7x^2 - 15$

17) $-4x^5 + 6x^4 - 3x^3 + 16$

18) $-2x^6 - 7x^2 - x$

19) $6x^7 + 8x^5 - 3x^2$

20) $3x^6 - 12x^5 + 10x^2$

21) $-11x^7 + 20x^5 - 6x^3$

22) $-5x^5 + 34x^4 + 6x^3$

23) $-6x^3 + 12x^2 + 15x$

24) $12x^7 + 48x^4$

25) $15x^3 + 30x^2 + 20x$

26) $42x^6 - 14x^2 + 28x$

27) $12x^5 - 12x^4 + 6x$

28) $8x^6 + 24x^3 - 12x$

29) $15x^5 + 20x^4 + 10x^2$

30) $8x^7 - 4x^4 + 6x^2$

Simplifying Polynomials

1) $12x - 60$

2) $15x^2 - 20x$

3) $30x^2 - 42x$

4) $21x^2 + 15x$

5) $20x^2 - 15x$

6) $48x^2 + 12x$

7) $3x^2 - 14x + 8$

8) $2x^2 - 4x - 30$

9) $x^2 - 10x + 21$

10) $9x^2 - 16$

11) $25x^2 - 30x + 8$

12) $-8x^4 + 12x^2$

13) $5x^3 - 2x^2 + 3x + 7$

14) $-10x^3 + 4x^2 + 7x$

15) $5x^5 - 6x^3 + 12x^2$

16) $6x^8 + 4x^6 - 5x^2$

17) $-4x^6 + 10x^5 - 12x^3 + 4x$

18) $-16x^3 - 3x^2 + 22$

19) $4x^3 + 2x^2 - 4x$

20) $-x^5 - 4x^2 + 18$

21) $x^6 - 14x^3$

22) $12x^4 - 4$

23) $-9x^4 - 12x^3$

24) -90

25) $-9x^4 - 2x + 19$

26) $-7x^5 - 2x^3 + 17$

27) $11x^3 - 10x + 2$

28) $-8x^4 + 2x^2$

29) $2x^5 + 6x + 2$

30) $11x^5 - 5x^4 + 4$

31) $-9x^5 - 28$

32) $9x^3 - 6x$

Adding and Subtracting Polynomials

1) $x^2 + 1$

2) $6x^3 - 1$

3) $2x^5 + 5x^2 - 15$

4) $6x^3 + 3x^2 - 4x$

5) $-24x^4 + 28x - 6$

6) $14x^2$

7) $14x^2 - 6$

8) $6x^2 - 4$

9) $-3x^3 + 2x$

10) $3x$

11) $19x^3 - 6x$

12) $-7x^5 - 4x$

13) $-11x^7 + 2x^2 + 6x$

14) $4x^5 + 11x^2 - 5$

15) $4x^5 + 18x^4$

16) $-9x^3 + 7x$

17) $-151x^2 + 3x$

18) $-11x^4 + 4x^2 + 5x$

19) $10x^4 - 3x^2 - 4$

20) $10x^6 + 6x^3$

21) $28x^5 + 17x^4 + 22x^2$

22) $10x^4 - 12x$

23) $32x^4 + 4x$

24) $5x^8 - 15x^6 - 5x^3 + 3x$

25) $6x^4 + 15x^2 - 12x$

26) $-2x^9 + 13x^7 + 40x^3$

Multiplying Monomials

1) $-6u^{10}$

2) $10p^{11}$

3) $12xy^3z^9$

4) $24u^6t^5$

5) $35a^5b^6$

6) $-18a^6b^4$

7) $13x^5y^9$

8) $-48p^5q^9$

9) $32s^5t^6$

10) $-36x^6y^4$

11) $36xy^7z^4$

12) $24px^3y^2$

13) $-39p^3q^5$

14) $13s^4t^8$

15) $-66p^8$

16) $-24p^4q^9r^7$

17) $28a^7b$

18) $-30u^9v^6$

19) $-27u^6$

20) $-24x^3y^6$

21) $-13y^8z^4$

22) $16a^5b^2c^6$

23) $35p^9q^8$

24) $-16u^{12}v^6$

25) $-17y^{10}z^6$

26) $-40p^3q^7r^3$

27) $15a^5b^6c^8$

28) $18x^5y^8z^5$

Multiplying and Dividing Monomials

1) $10x^7$

2) $24x^6$

3) $21x^8$

4) $20x^8$

5) $36x^{10}$

6) $32x^{11}y^5$

7) $14x^7y^6$

8) $-10x^6y^9$

9) $18x^7y^7$

10) $-25x^5y^4$

11) $24x^7y^7$

12) $20x^5y^6$

13) $48x^7y^{16}$

14) $45x^7y^{11}$

15) $56x^8y^{14}$

16) $-21x^{12}y^{12}$

17) $5x^5y^2$

18) xy^4

19) $7x^3y^3$

20) $9xy$

21) $4x^5y^3$

22) $12x^5y$

23) $8x^{10}y^3$

24) $5x^{-1}y^3$

25) $5x^3$

26) $-7x^{14}y^5$

27) $-4x^2$

Multiplying a Polynomial and a Monomial

1) $2x^2 + 4x$

2) $-12x + 24$

3) $20x^2 + 10x$

4) $-4x^2 + 5x$

5) $16x^2 - 16x$

6) $12x - 24y$

7) $35x^2 - 35x$

8) $36x^2 + 6xy$

9) $4x^2 + 24xy$

10) $33x^2 + 44xy$

11) $21x^2 + 14x$

12) $40x^2 - 100xy$

13) $27x^2 - 18xy$

14) $7x^2 - 28xy + 42x$

15) $16x^3 + 40xy^2$

16) $24x^2 + 36xy$

17) $8x^4 - 16y^4$

18) $-12x^3y + 16xy$

19) $-20x^3 + 8xy - 16$

20) $4x^2 - 20xy - 24$

21) $16x^4 - 40x^2y + 16x^2$

22) $12x^4 + 36x^2 - 12x^2y$

23) $6x^2 + 3xy - 27y^2$

24) $20x^4 - 12x^2 + 28x$

25) $18x^{22} - 12x - 30$

26) $-2x^5 + 4x^3 + 3x^2$

27) $4x^5 - 2x^3 + 10x^2$

28) $12x^7 - 8x^5 + 20x^4$

29) $8x^6 - 10x^3y + 14x^2y^3$

30) $25x^6 - 15x^3 + 45x^2$

31) $42x^4 + 21x^3 - 42x^2$

32) $4x^4 - 16x^2y + 8xy^2$

Multiplying Binomials

1) $x^2 + 9x + 18$

2) $x^2 - x - 12$

3) $x^2 - 11x + 24$

4) $x^2 + 17x + 72$

5) $x^2 - 14x + 24$

6) $x^2 + 10x + 25$

7) $x^2 + x - 42$

8) $x^2 - 11x + 24$

9) $x^2 + 19x + 84$

10) $x^2 + 4x - 32$

11) $x^2 + 16x + 64$

12) $x^2 + 9x + 14$

13) $x^2 - 36$

14) $x^2 - 25$

15) $x^2 + 22x + 121$

16) $x^2 + 15x + 54$

17) $x^2 - 4$

18) $x^2 + 3x - 28$

19) $3x^2 + 23x + 30$

20) $20x^2 + 16x - 48$

21) $3x^2 - 14x - 49$

22) $x^2 - 13x + 36$

23) $x^2 - 10x - 24$

24) $10x^2 - 12x - 16$

25) $3x^2 + 16x - 64$

26) $42x^2 + 9x - 6$

27) $12x^2 + 35x + 25$

28) $63x^2 - 8x - 16$

29) $2x^2 - 4x - 16$

30) $25x^2 - 16$

31) $9x^2 - 15x - 14$

32) $x^4 - 64$

Factoring Trinomials

1) $(x + 6)(x + 2)$

2) $(x - 5)(x - 1)$

3) $(x + 12)(x + 3)$

4) $(x - 5)(x - 7)$

5) $(x - 2)(x - 9)$

6) $(x - 6)(x - 3)$

7) $(x + 6)(x + 12)$

8) $(x + 8)(x - 9)$

9) $(x - 3)(x + 7)$

10) $(x - 11)(x - 2)$

11) $(x - 4)(x + 6)$

12) $(x - 8)(x + 5)$

13) $(x + 7)(x - 10)$

14) $(x + 13)(x + 13)$

15) $(4x + 5)(x - 3)$

16) $(x - 11)(x - 3)$

17) $(5x - 5)(2x + 3)$

18) $(2x - 6)(3x + 7)$

19) $(x + 6)(x + 6)$

20) $(5x - 3)(x + 4)$

21) $(x - 8)$

22) $(4x - 3)$

23) $(2x - 6)$

Operations with Polynomials

1) $20x + 12$

2) $16x + 48$

3) $10x - 4$

4) $-28x + 12$

5) $27x^3 + 3x^2$

6) $28x^7 - 36x^6$

7) $-21x^5 + 9x^4$

8) $-40x^5 + 64x^4$

9) $7x^2 + 35x - 21$

10) $45x^2 - 63x + 45$

11) $9x^2 + 9x + 6$

12) $15x^3 + 25x^2 + 40x$

13) $15x^2 + 6x - 21$

14) $27x^2 - 36x - 15$

15) $24x^2 - 6$

16) $21x^2 + 29x - 10$

17) $(5x + 2y)$

18) $16x^2 - 2x - \frac{15}{2}$

19) $36x^2 + 24x + 4$

20) $40x$

21) $x^3 + 6x^2 + 12x + 8$

22) $(9x + y)$

Chapter 9 :

Functions Operations and Quadratic

Topics that you will practice in this chapter:

- ✓ Evaluating Function
- ✓ Adding and Subtracting Functions
- ✓ Multiplying and Dividing Functions
- ✓ Composition of Functions
- ✓ Quadratic Equation
- ✓ Solving Quadratic Equations
- ✓ Quadratic Formula and the Discriminant
- ✓ Quadratic Inequalities
- ✓ Graphing Quadratic Functions
- ✓ Domain and Range of Radical Functions
- ✓ Solving Radical Equations

It's fine to work on any problem, so long as it generates interesting mathematics along the way – even if you don't solve it at the end of the day." – Andrew Wiles

Evaluating Function

✍ **Write each of following in function notation.**

1) $h = -8x + 3$

2) $k = 2a - 14$

3) $d = 11t$

4) $y = \frac{5}{12}x - \frac{7}{12}$

5) $m = 24n - 210$

6) $c = p^2 - 5p + 10$

✍ **Evaluate each function.**

7) $f(x) = 2x - 7$, find $f(-3)$

8) $g(x) = \frac{1}{9}x + 12$, find $f(18)$

9) $h(x) = -4x + 9$, find $f(3)$

10) $f(x) = -x + 19$, find $f(-3)$

11) $f(a) = 7a - 12$, find $f(3)$

12) $h(x) = 14 - 3x$, find $f(-4)$

13) $g(n) = 6n - 10$, find $f(2)$

14) $f(x) = -11x - 4$, find $f(-1)$

15) $k(n) = -20 - 3.5n$, find $f(2)$

16) $f(x) = -0.7x + 3.3$, find $f(-7)$

17) $g(n) = \frac{11n+8}{n}$, find $g(2)$

18) $g(n) = \sqrt{3n} + 12$, find $g(3)$

19) $h(x) = x^{-2} - 7$, find $h(\frac{1}{9})$

20) $h(n) = n^{-3} + 11$, find $h(\frac{1}{4})$

21) $h(n) = n^3 - 2$, find $h(\frac{1}{2})$

22) $h(n) = n^2 - 4$, find $h(-\frac{1}{3})$

23) $h(n) = 4n^2 - 13$, find $h(-5)$

24) $h(n) = -2n^3 - 6n$, find $h(2)$

25) $g(n) = \sqrt{16n^2} - \sqrt{n}$, find $g(4)$

26) $h(a) = \frac{-14a+9}{3a}$, find $h(-b)$

27) $k(a) = 12a - 14$, find $k(a - 3)$

28) $h(x) = \frac{1}{9}x + 18$, find $h(-18x)$

29) $h(x) = 8x^2 + 16$, find $h(\frac{x}{2})$

30) $h(x) = x^4 - 20$, find $h(-2x)$

Adding and Subtracting Functions

✍ **Perform the indicated operation.**

1) $f(x) = 2x + 3$

   $g(x) = x + 7$

   Find $(f - g)(2)$

2) $g(a) = -5a - 8$

   $f(a) = -3a - 5$

   Find $(g - f)(-2)$

3) $h(t) = 4t + 3$

   $g(t) = 4t + 7$

   Find $(h - g)(t)$

4) $g(a) = -6a - 10$

   $f(a) = 3a^2 + 9$

   Find $(g - f)(x)$

5) $g(x) = \frac{5}{6}x - 23$

   $h(x) = \frac{5}{12}x + 25$

   Find $g(12) - h(12)$

6) $h(x) = \sqrt{3x} - 2$

   $g(x) = \sqrt{3x} + 5$

   Find $(h + g)(12)$

7) $f(x) = x^{-1}$

   $g(x) = x^2 + \frac{5}{x}$

   Find $(f - g)(-3)$

8) $h(n) = n^2 + 2$

   $g(n) = -4n + 6$

   Find $(h - g)(2a)$

9) $g(x) = -2x^2 - 5 - 4x$

   $f(x) = 7 + 2x$

   Find $(g - f)(3x)$

10) $g(t) = 11t - 4$

    $f(t) = -2t^2 + 5$

    Find $(g + f)(-t)$

11) $f(x) = 8x + 9$

    $g(x) = -5x^2 + 3x$

    Find $(f - g)(-x^2)$

12) $f(x) = -3x^4 - 5x$

    $g(x) = 2x^4 + 5x + 22$

    Find $(f + g)(3x^2)$

Multiplying and Dividing Functions

✎ **Perform the indicated operation.**

1) $g(x) = -2x - 1$

    $f(x) = 4x + 3$

    Find $(g.f)(2)$

2) $f(x) = 5x$

    $h(x) = -2x + 3$

    Find $(f.h)(-2)$

3) $g(a) = 5a - 2$

    $h(a) = 2a - 3$

    Find $(g.h)(-3)$

4) $f(x) = 2x - 7$

    $h(x) = x - 5$

    Find $(\frac{f}{h})(4)$

5) $f(x) = 8a^2$

    $g(x) = 3 + 2a$

    Find $(\frac{f}{g})(2)$

6) $g(a) = \sqrt{4a} + 2$

    $f(a) = (-a)^4 + 1$

    Find $(\frac{g}{f})(1)$

7) $g(t) = t^3 + 1$

    $h(t) = 5t - 2$

    Find $(g.h)(-2)$

8) $g(n) = n^2 + 2n - 4$

    $h(n) = -5n + 3$

    Find $(g.h)(1)$

9) $g(a) = (a - 3)^2$

    $f(a) = a^2 + 4$

    Find $(\frac{g}{f})(3)$

10) $g(x) = -3x^2 + \frac{4}{5}x + 9$

    $f(x) = x^2 - 24$

    Find $(\frac{g}{f})(5)$

11) $f(x) = 2x^3 - 5x^2 + 1$

    $g(x) = 3x - 1$

    Find $(f.g)(x)$

12) $f(x) = 5x - 2$

    $g(x) = x^3 - 2x$

    Find $(f.g)(x^2)$

Composition of Functions

✎ Using $f(x) = 2x - 5$ and $g(x) = -2x$, find:

1) $f(g(2)) =$

2) $f(g(-1)) =$

3) $g(f(-4)) =$

4) $g(f(5)) =$

5) $f(g(3)) =$

6) $g(f(0)) =$

✎ Using $f(x) = -\frac{1}{4}x + \frac{3}{4}$ and $g(x) = 2x^2$, find:

7) $g(f(-2)) =$

8) $g(f(4)) =$

9) $g(g(1)) =$

10) $f(f(1)) =$

11) $g(f(-4)) =$

12) $f(g(x)) =$

✎ Using $f(x) = -2x + 2$ and $g(x) = x + 1$, find:

13) $g(f(1)) =$

14) $f(f(0)) =$

15) $f(g(-1)) =$

16) $f(g(-3)) =$

17) $g(f(2)) =$

18) $f(g(x)) =$

✎ Using $f(x) = \sqrt{x + 9}$ and $g(x) = x - 9$, find:

19) $f(g(9)) =$

20) $g(f(-9)) =$

21) $f(g(4)) =$

22) $f(f(7)) =$

23) $g(f(-5)) =$

24) $g(g(0)) =$

Quadratic Equation

✎ Multiply.

1) $(x - 4)(x + 6) = $ _____

2) $(x + 5)(x + 7) = $ _____

3) $(x - 6)(x + 8) = $ _____

4) $(x + 2)(x - 9) = $ _____

5) $(x - 7)(x - 8) = $ _____

6) $(3x + 2)(x - 3) = $ _____

7) $(4x - 3)(x + 2) = $ _____

8) $(4x - 5)(x + 1) = $ _____

9) $(7x + 1)(x - 6) = $ _____

10) $(5x + 1)(3x - 3) = $ _____

✎ Factor each expression.

11) $x^2 - 2x - 8 = $ _____

12) $x^2 + 8x + 15 = $ _____

13) $x^2 - 2x - 24 = $ _____

14) $x^2 - 10x + 21 = $ _____

15) $x^2 + 10x + 21 = $ _____

16) $4x^2 + 9x + 5 = $ _____

17) $5x^2 + 13x - 6 = $ _____

18) $5x^2 + 17x - 12 = $ _____

19) $2x^2 + 7x + 5 = $ _____

20) $9x^2 - 21x + 6 = $ _____

✎ Calculate each equation.

21) $(x + 6)(x - 3) = 0$

22) $(x + 1)(x + 8) = 0$

23) $(3x + 6)(x + 5) = 0$

24) $(2x - 2)(4x + 8) = 0$

25) $x^2 + x + 10 = 22$

26) $x^2 + 11x + 36 = 12$

27) $2x^2 + 9x + 9 = 5$

28) $x^2 + 3x - 24 = 4$

29) $5x^2 + 5x - 40 = 20$

30) $8x^2 + 8x = 48$

Solving Quadratic Equations

✎ **Solve each equation by factoring or using the quadratic formula.**

1) $(x + 9)(x - 1) = 0$

2) $(x + 7)(x + 6) = 0$

3) $(x - 8)(x + 3) = 0$

4) $(x - 6)(x - 4) = 0$

5) $(x + 2)(x + 12) = 0$

6) $(5x + 4)(x + 7) = 0$

7) $(6x + 1)(4x + 5) = 0$

8) $(2x + 7)(x + 8) = 0$

9) $(x + 6)(3x + 15) = 0$

10) $(12x + 2)(x + 8) = 0$

11) $x^2 = 8x$

12) $x^2 - 16 = 0$

13) $3x^2 + 6 = 9x$

14) $-2x^2 - 8 = 10x$

15) $5x^2 + 40x = 45$

16) $x^2 + 10x = 24$

17) $x^2 + 6x = 16$

18) $x^2 + 9x = -18$

19) $x^2 + 13x = -36$

20) $x^2 + 3x - 15 = 5x$

21) $x^2 + 8x + 7 = -8$

22) $3x^2 - 11x = -9 + x$

23) $10x^2 + 3 = 27x - 15$

24) $7x^2 - 6x + 8 = 8$

25) $2x^2 - 12 = -3x + 2$

26) $10x^2 - 26x - 3 = -15$

27) $3x^2 + 21 = -16x + 5$

28) $x^2 + 15x - 10 = -66$

29) $3x^2 - 8x - 8 = 4 + x$

30) $2x^2 + 6x - 24 = 12$

31) $3x^2 - 33x + 54 = -18$

32) $-10x^2 - 15x - 9 = -9 - 27x^2$

Quadratic Formula and the Discriminant

✎ Find the value of the discriminant of each quadratic equation.

1) $3x(x - 8) = 0$

2) $2x^2 + 6x - 4 = 0$

3) $x^2 + 6x + 7 = 0$

4) $x^2 - x + 3 = 0$

5) $x^2 + 4x - 3 = 0$

6) $2x^2 + 6x - 10 = 0$

7) $3x^2 + 7x + 5 = 0$

8) $x^2 - 6x - 4 = 0$

9) $2x^2 + 8x + 3 = 0$

10) $x^2 + 7x - 5 = 0$

11) $5x^2 + 2x - 3 = 0$

12) $-3x^2 - 11x + 4 = 0$

13) $-6x^2 - 12x + 8 = 0$

14) $-x^2 - 9x - 12 = 0$

15) $7x^2 - 6x - 10 = 0$

16) $-4x^2 - 2x + 8 = 0$

17) $5x^2 + 8x - 2 = 0$

18) $6x^2 - 4x = 0$

19) $3x^2 - 5x + 2 = 0$

20) $4x^2 + 9x + 3 = 0$

✎ Find the discriminant of each quadratic equation then state the number of real and imaginary solutions.

21) $-4x^2 - 16 = 16x$

22) $20x^2 = 20x - 5$

23) $-11x^2 - 19x = 26$

24) $22x^2 - 4x + 1 = 18x^2$

25) $-11x^2 = -15x + 8$

26) $3x^2 + 6x + 9 = 6$

27) $13x^2 - 5x - 12 = -26$

28) $-8x^2 - 32x - 25 = 7$

Graphing Quadratic Functions

✎ketch the graph of each function. Identify the vertex and axis of symmetry.

1) $y = (x + 3)^2 + 2$

2) $y = (x - 3)^2 - 2$

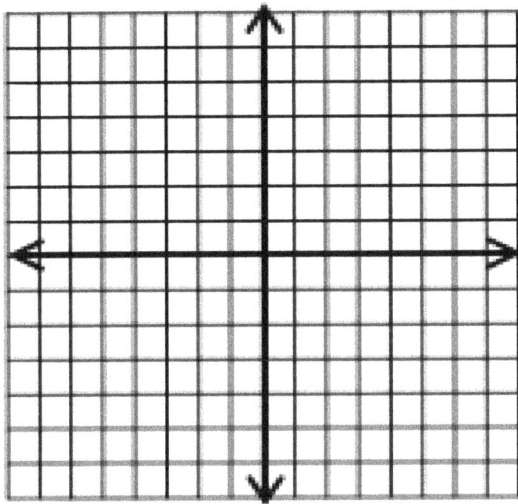

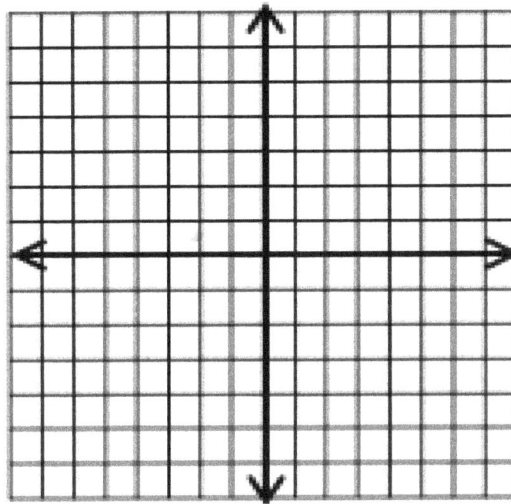

3) $y = 6 - (-x + 4)^2$

4) $y = -3x^2 - 6x + 9$

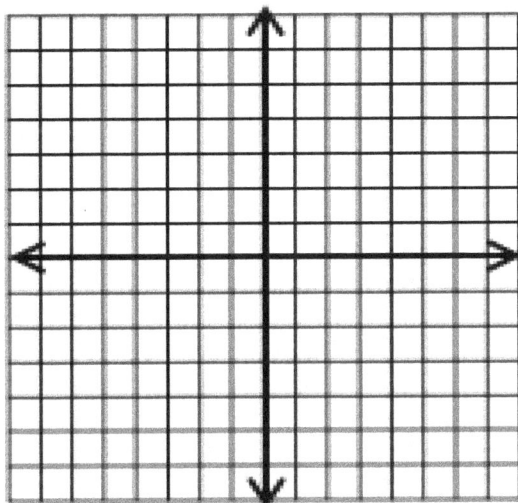

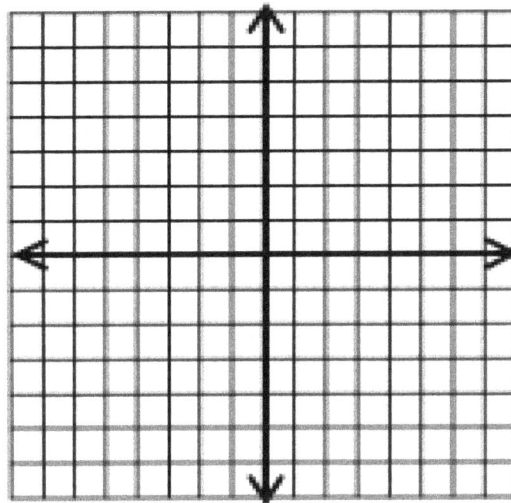

Quadratic Inequalities

✎ **Solve each quadratic inequality.**

1) $x^2 - 25 < 0$

2) $-x^2 - 6x - 8 > 0$

3) $5x^2 + 15x + 30 < 0$

4) $x^2 + 8x + 16 > 0$

5) $2x^2 - 18x - 20 \geq 0$

6) $x^2 > -10x - 25$

7) $3x^2 + 2x + 16 \leq 0$

8) $x^2 - 5x - 14 \leq 0$

9) $x^2 - 6x - 7 \geq 0$

10) $2x^2 + 16x - 18 < 0$

11) $x^2 + 6x - 72 > 0$

12) $3x^2 - 3x - 36 > 0$

13) $x^2 - 15x + 64 \leq 0$

14) $2x^2 - 24x + 72 \leq 0$

15) $x^2 - 16x + 63 \geq 0$

16) $x^2 - 16x + 55 \geq 0$

17) $x^2 - 81 \leq 0$

18) $x^2 - 17x + 42 \geq 0$

19) $9x^2 + 14x + 36 \leq 0$

20) $4x^2 - 2x - 24 > 2x^2$

21) $5x^2 - 20x + 20 < 0$

22) $7x^2 - 6x \geq 6x^2 - 5$

23) $5x^2 - 15 > 4x^2 + 2x$

24) $3x^2 - 4x \geq 3x^2 - 9x + 15$

25) $8x^2 + 9x - 54 > 5x^2$

26) $10x^2 + 50x - 60 < 0$

27) $-x^2 + 15x - 57 \geq 0$

28) $-5x^2 + 25x + 30 \leq 0$

29) $5x^2 + 40x + 75 < 0$

30) $9x^2 + 20x + 180 \leq 0$

31) $3x^2 + 2x - 36 \geq -x$

32) $3x^2 + 9x + 9 \leq 6x^2 + 3x$

Domain and Range of Radical Functions

✏️ **Identify the domain and range of each function.**

1) $y = \sqrt{x+8} - 7$

2) $y = \sqrt[3]{3x-5} - 4$

3) $y = \sqrt{3x-9} + 3$

4) $y = \sqrt[3]{(4x+6)} - 2$

5) $y = 3\sqrt{4x+20} + 6$

6) $y = \sqrt[3]{(5x-2)} - 11$

7) $y = 4\sqrt{9x^2+8} + 3$

8) $y = \sqrt[3]{(7x^2-2)} - 6$

9) $y = 2\sqrt{2x^3+16} - 3$

10) $y = \sqrt[3]{(11x+4)} - 2x$

11) $y = 3\sqrt{-2(4x+8)} + 5$

12) $y = \sqrt[5]{(3x^2-12)} - 6$

13) $y = 3\sqrt{x-5} - 2$

14) $y = \sqrt[3]{6x+9} - 4$

✏️ **Sketch the graph of each function.**

15) $y = -3\sqrt{x} + 5$

16) $y = 3\sqrt{x} - 6$

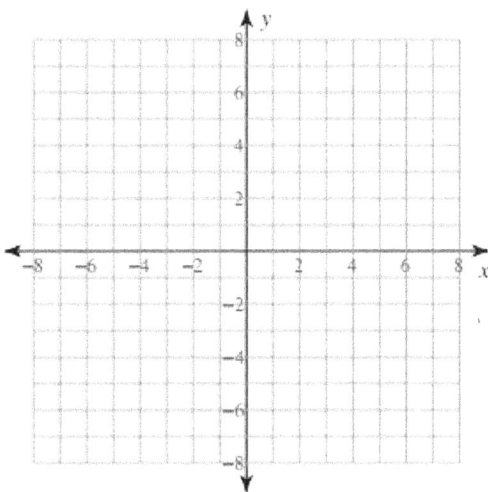

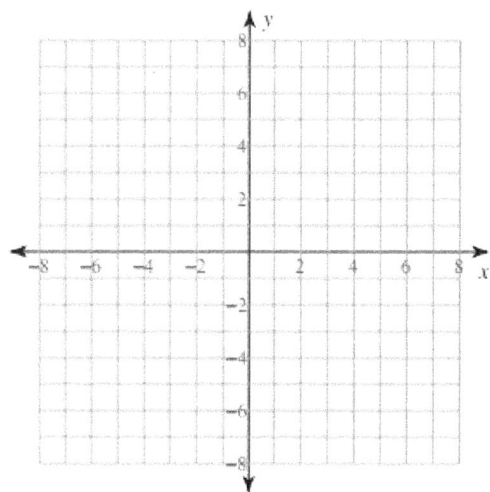

Solving Radical Equations

✑ **Solve each equation. Remember to check for extraneous solutions.**

1) $\sqrt{a} = 9$

2) $\sqrt{v} = 6$

3) $\sqrt{r} = 4$

4) $8 = 16\sqrt{x}$

5) $\sqrt{x + 3} = 18$

6) $6 = \sqrt{x - 7}$

7) $4 = \sqrt{r - 3}$

8) $\sqrt{x - 5} = 7$

9) $12 = \sqrt{x - 4}$

10) $\sqrt{m + 5} = 8$

11) $7\sqrt{5a} = 35$

12) $6\sqrt{2x} = 48$

13) $2 = \sqrt{6x - 32}$

14) $\sqrt{304 - 4x} = 4$

15) $\sqrt{r + 2} - 8 = 6$

16) $-21 = -7\sqrt{r + 9}$

17) $60 = 6\sqrt{5v}$

18) $x = \sqrt{40 - 3x}$

19) $\sqrt{90 - 27a} = 3a$

20) $\sqrt{-8n + 88} = 4$

21) $\sqrt{15r - 5} = 4r - 3$

22) $\sqrt{-64 + 32x} = 4x$

23) $\sqrt{4x + 15} = \sqrt{2x + 11}$

24) $\sqrt{12v} = \sqrt{15v - 21}$

25) $\sqrt{9 - x} = \sqrt{x - 3}$

26) $\sqrt{6m + 34} = \sqrt{8m + 34}$

27) $\sqrt{7r + 32} = \sqrt{-8 - 3r}$

28) $\sqrt{4k + 10} = \sqrt{2 - 4k}$

29) $-20\sqrt{x - 13} = -40$

30) $\sqrt{90 - 2x} = \sqrt{\dfrac{x}{4}}$

Answers of Worksheets

Evaluating Function

1) $h(x) = -8x + 3$

2) $k(a) = 2a - 14$

3) $d(t) = 11t$

4) $f(x) = \frac{5}{12}x - \frac{7}{12}$

5) $m(n) = 24n - 210$

6) $c(p) = p^2 - 5p + 10$

7) -13

8) 14

9) -3

10) 22

11) 9

12) 26

13) 2

14) 7

15) -27

16) 8.2

17) 15

18) 15

19) 74

20) 75

21) $-1\frac{7}{8}$

22) $-3\frac{8}{9}$

23) 87

24) -28

25) 14

26) $-\frac{14b+9}{3b}$

27) $12a - 50$

28) $-2x + 18$

29) $2x^2 + 16$

30) $16x^4 - 20$

Adding and Subtracting Functions

1) -2

2) 1

3) -4

4) $-3x^2 - 6x - 19$

5) -43

6) 15

7) $-7\frac{2}{3}$

8) $4a^2 + 8a - 4$

9) $-18x^2 - 18x - 12$

10) $-2t^2 - 11t + 1$

11) $5x^4 - 5x^2 + 9$

12) $-81x^8 + 22$

Multiplying and Dividing Functions

1) -55

2) -70

3) 153

4) -1

5) $4\frac{4}{7}$

6) 2

7) 84

8) 2

9) 0

10) -62

11) $6x^4 - 17x^3 + 5x^2 + 3x - 1$

12) $5x^8 - 2x^6 - 10x^4 + 4x^2$

Composition of Functions

1) -13

2) -1

3) 26

4) -10

5) -17

6) 10

7) $\frac{25}{8}$

8) $\frac{1}{8}$

9) 8

10) $\frac{5}{8}$

11) $\frac{49}{8}$

12) $-\frac{1}{2}(x^2 - \frac{3}{2})$

13) 1

14) -2

15) 2

16) 6

17) -1

18) $-2x$

19) 3   21) 2   23) -7

20) -9 22) $\sqrt{13}$ 24) -18

Quadratic Equations

1) $x^2 + 2x - 24$
2) $x^2 + 12x + 35$
3) $x^2 + 2x - 48$
4) $x^2 - 7x - 18$
5) $x^2 - 15x + 56$
6) $3x^2 - 7x - 6$
7) $4x^2 + 5x - 6$
8) $4x^2 - x - 5$
9) $7x^2 - 41x - 6$
10) $15x^2 - 12x - 3$

11) $(x - 4)(x + 2)$
12) $(x + 5)(x + 3)$
13) $(x - 6)(x + 4)$
14) $(x - 3)(x - 7)$
15) $(x + 3)(x + 7)$
16) $(4x + 5)(x + 1)$
17) $(5x - 2)(x + 3)$
18) $(5x - 3)(x + 4)$
19) $(2x + 5)(x + 1)$
20) $3(x - 2)(3x - 1)$

21) $x = -6, x = 3$
22) $x = -1, x = -8$
23) $x = -2, x = -5$
24) $x = 1, x = -2$
25) $x = 3, x = -4$
26) $x = -3, x = -8$
27) $x = -4, x = -\frac{1}{2}$
28) $x = 4, x = -7$
29) $x = 3, x = -4$
30) $x = -3, x = 2$

Solving quadratic equations

1) $\{-9, 1\}$
2) $\{-6, -7\}$
3) $\{8, -3\}$
4) $\{6, 4\}$
5) $\{-2, -12\}$
6) $\{-\frac{4}{5}, -7\}$
7) $\{-\frac{5}{4}, -\frac{1}{6}\}$
8) $\{-\frac{7}{2}, -8\}$

9) $\{-6, -5\}$
10) $\{-\frac{1}{6}, -8\}$
11) $\{8, 0\}$
12) $\{4, -4\}$
13) $\{2, 1\}$
14) $\{-4, -1\}$
15) $\{1, -9\}$
16) $\{2, -12\}$

17) $\{2, -8\}$
18) $\{-3, -6\}$
19) $\{-4, -9\}$
20) $\{5, -3\}$
21) $\{-5, -3\}$
22) $\{1, 3\}$
23) $\{\frac{6}{5}, \frac{3}{2}\}$
24) $\{\frac{6}{7}, 0\}$

25) $\{-\frac{7}{2}, 2\}$
26) $\{\frac{3}{5}, 2\}$
27) $\{-\frac{4}{3}, -4\}$
28) $\{-8, -7\}$
29) $\{4, -1\}$
30) $\{3, -6\}$
31) $\{3, 8\}$
32) $\{\frac{15}{17}, 0\}$

Quadratic formula and the discriminant

1) 576
2) 68
3) 8
4) -11
5) 28

6) 116
7) -11
8) 52
9) 40
10) 69

11) 64
12) 169
13) 336
14) 33
15) 316

16) 132
17) 104
18) 16
19) 1
20) 33

21) 0, *one real solution* 22) 0, *one real solution* 23) -783, *no solution*

24) 0, *one real solution*   26) 0, *one real solution*   28) 0, *one real solution*

25) -127, *no solution*   27) -703, *no solution*

Graphing quadratic functions

1) $(-3, 2), x = -3$ 2) $(3, -2), x = 3$

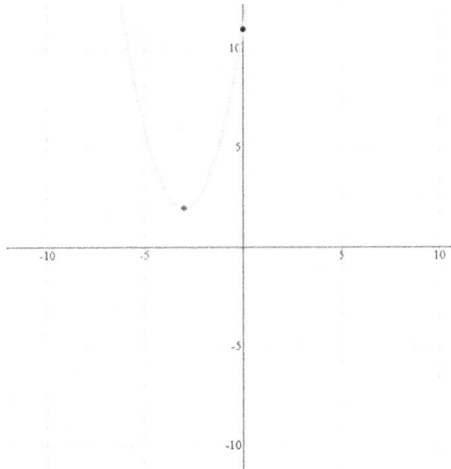

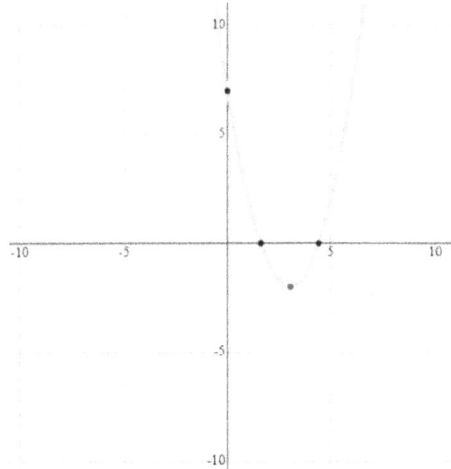

3) $(4, 6), x = 4$ 4) $(-1, 12), x = -1$

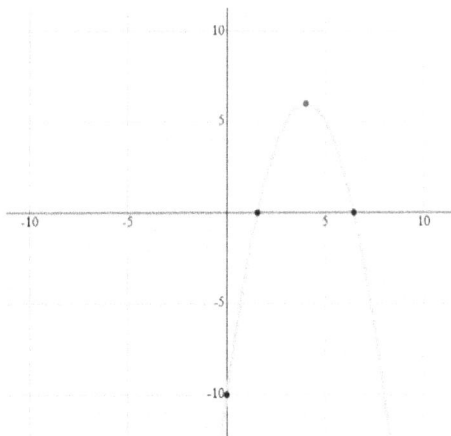

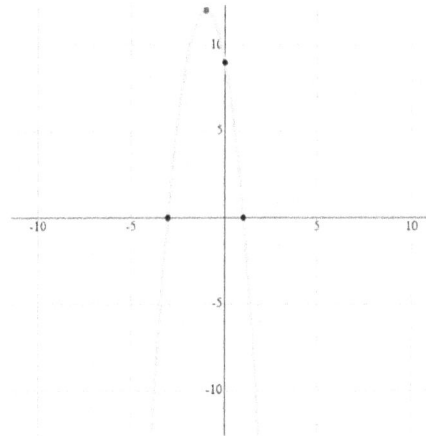

Quadratic inequalities

1) $-5 < x < 5$ 7) no solution   13) no solution

2) $-4 < x < -2$ 8) $-2 \leq x \leq 7$ 14) $x = 6$

3) no solution   9) $x \leq -1 \ or \ x \geq 7$ 15) $x \leq 7 or \ x \geq 9$

4) $x < -4 \ or \ x > -4$ 10) $-9 < x < 1$ 16) $x \leq 5 or \ x \geq 11$

5) $x \leq -1 \ or \ x \geq 10$ 11) $x < -12 \ or \ x > 6$ 17) $-9 \leq x \leq 9$

6) $x < -5 \ or \ x > -5$ 12) $-3 < x < 4$ 18) $x \leq 3 \ or \ x \geq 14$

19) no solution

20) $x < -3$ or $x > 4$

21) no solution

22) $x \leq 1$ or $x \geq 5$

23) $x < -3$ or $x > 5$

24) $x \geq 3$

25) $x < -6$ or $x > 3$

26) $-6 < x < 1$

27) no solution

28) $x \leq -1$ or $x \geq 6$

29) $-5 < x < -3$

30) no solution

31) $x \leq -4$ or $x \geq 3$

32) $x \leq -1$ or $x \geq 3$

Domain and range of radical functions

1) domain: $x \geq -8$

   range: $y \geq -7$

2) domain: {all real numbers}

   range: {all real numbers}

3) domain: $x \geq 3$

   range: $y \geq 3$

4) domain: {all real numbers}

   range: {all real numbers}

5) domain: $x \geq -5$

   range: $y \geq 6$

6) domain: {all real numbers}

   range: {all real numbers}

7) domain: {all real numbers}

   range: $y \geq 8\sqrt{2} + 3$

8) domain: {all real numbers}

   range: {all real numbers}

9) domain: $x \geq -2$

   range: $y \geq -3$

10) domain: {all real numbers}

    range: {all real numbers}

11) domain: $x \leq -2$

    range: $y \geq 5$

12) domain: {all real numbers}

    range: {all real numbers}

13) domain: $x \geq 5$

    range: $y \geq -2$

14) domain: {all real numbers}

    range: {all real numbers}

15)

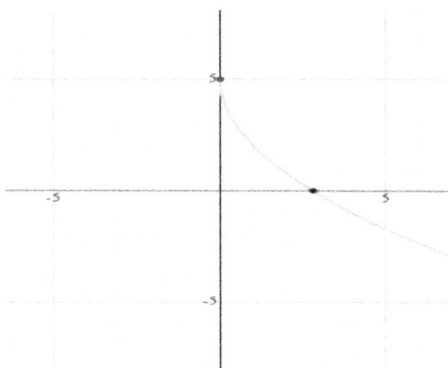

16)

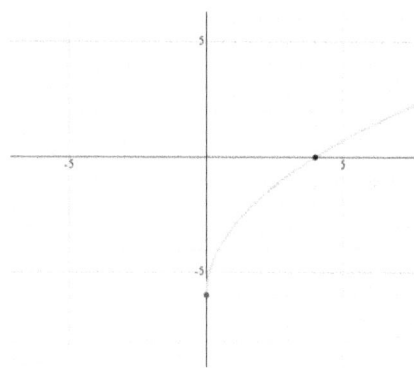

Solving radical equations

1) {81}

2) {36}

3) {16}

4) $\{\frac{1}{4}\}$

5) {321}

6) {43}

7) {19}

8) {54}

9) {148}

10) {59}

11) {5}

12) {32}

13) {6}

14) {72}

15) {194}

16) {0}

17) {20}

18) {5}

19) {2}

20) {9}

21) {2}

22) no solution

23) {−2}

24) {7}

25) {6}

26) {0}

27) {−4}

28) {−1}

29) {17}

30) {40}

Chapter 10 :

Geometry and Solid Figures

Topics that you will practice in this chapter:

✓ Angles

✓ Pythagorean Relationship

✓ Triangles

✓ Polygons

✓ Trapezoids

✓ Circles

✓ Cubes

✓ Rectangular Prism

✓ Cylinder

✓ Pyramids and Cone

Mathematics is, as it were, a sensuous logic, and relates to philosophy as do the arts, music, and plastic art to poetry. — K. Shegel

Angles

🖎 **What is the value of x in the following figures?**

1)

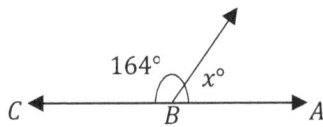

2)

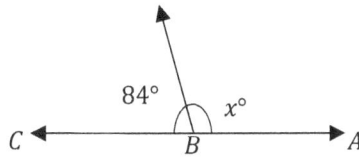

3)

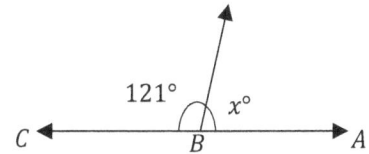

4)

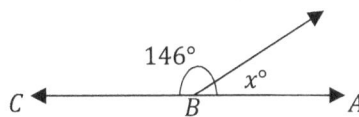

5)

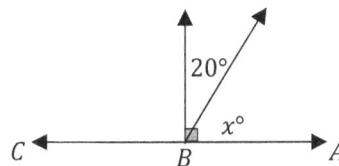

6)

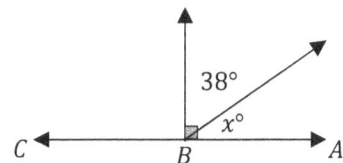

🖎 **Calculate.**

7) Two supplement angles have equal measures. What is the measure of each angle? _____

8) The measure of an angle is seven fifth the measure of its supplement. What is the measure of the angle? _____

9) Two angles are complementary and the measure of one angle is 24 less than the other. What is the measure of the smaller angle? _____

10) Two angles are complementary. The measure of one angle is one fifth the measure of the other. What is the measure of the bigger angle? _____

11) Two supplementary angles are given. The measure of one angle is 40° less than the measure of the other. What does the smaller angle measure? _____

Pythagorean Relationship

✎ **Do the following lengths form a right triangle?**

1)

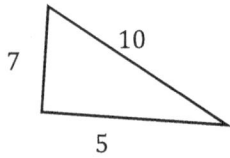
7, 10, 5

2)

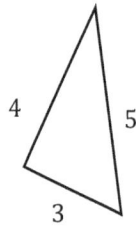
4, 5, 3

3)

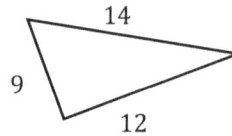
14, 9, 12

4)

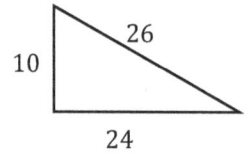
10, 26, 24

5)

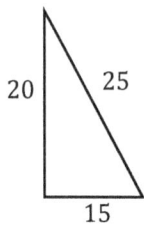
20, 25, 15

6)

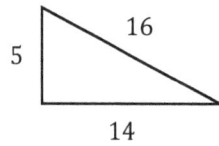
5, 16, 14

7)

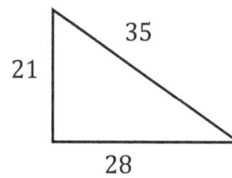
21, 35, 28

8)

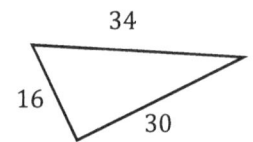
16, 34, 30

✎ **Find the missing side?**

9)

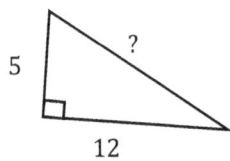
5, ?, 12

10)

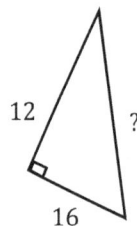
12, ?, 16

11)

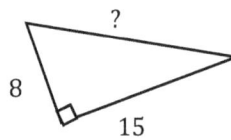
8, ?, 15

12)

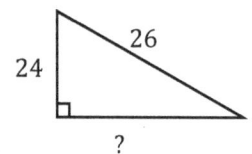
24, 26, ?

13)

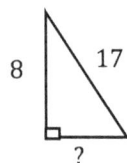
8, 17, ?

14)

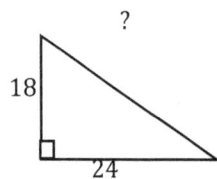
18, ?, 24

15)

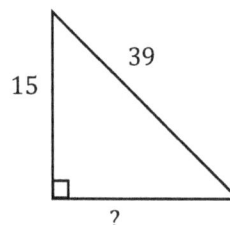
15, 39, ?

16)

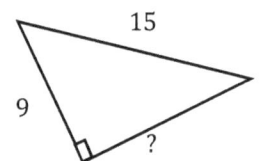
9, 15, ?

Triangles

✎ **Find the measure of the unknown angle in each triangle.**

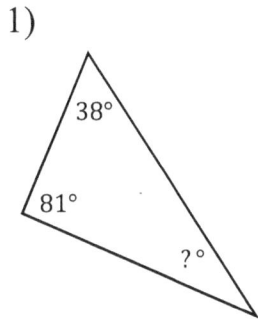

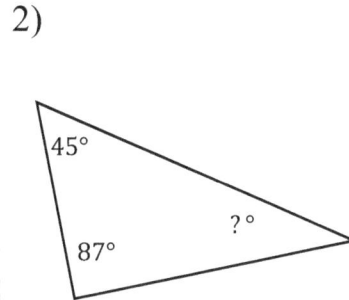

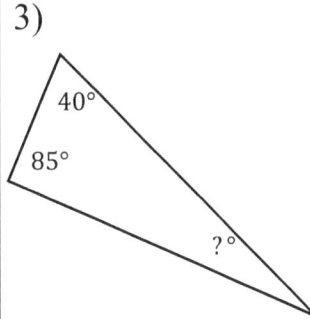

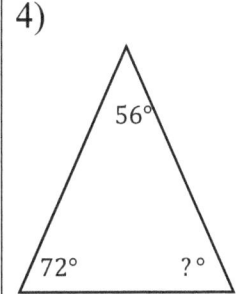

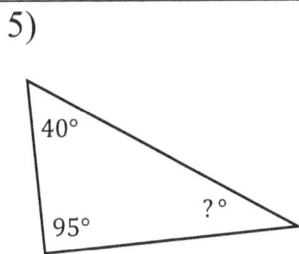

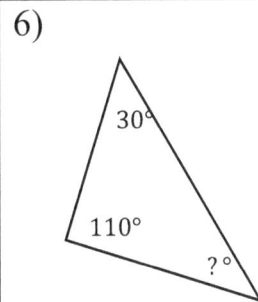

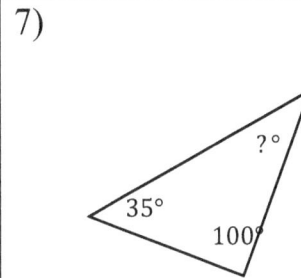

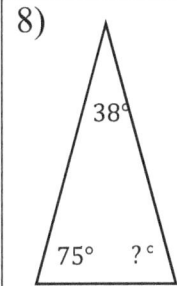

1)

38°
81°
?°

2)

45°
87°
?°

3)

40°
85°
?°

4)

56°
72°
?°

5)

40°
95°
?°

6)

30°
110°
?°

7)

?°
35°
100°

8)

38°
75°
?°

✎ **Find area of each triangle.**

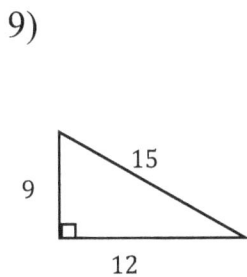

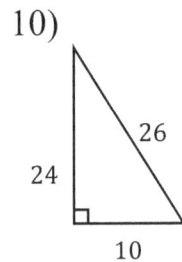

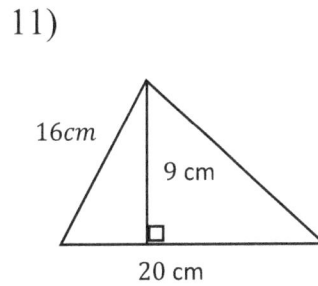

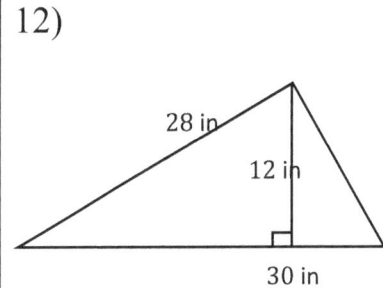

9)

9
15
12

10)

24
26
10

11)

16cm
9 cm
20 cm

12)

28 in
12 in
30 in

Polygons

✎ **Find the perimeter of each shape.**

1)

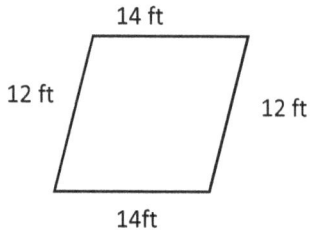

14 ft
12 ft   12 ft
14ft

2)

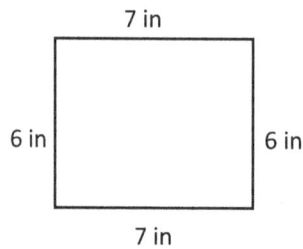

7 in
6 in   6 in
7 in

3)

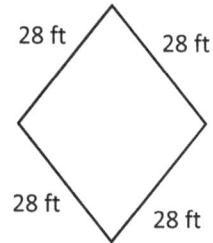

28 ft   28 ft
28 ft   28 ft

4) Square

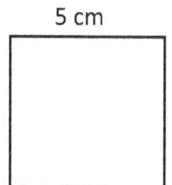

5 cm

5) Regular hexagon

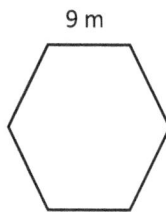

9 m

6)

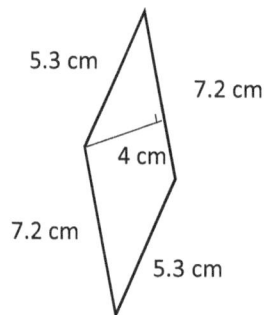

5.3 cm
7.2 cm
4 cm
7.2 cm
5.3 cm

7) Parallelogram

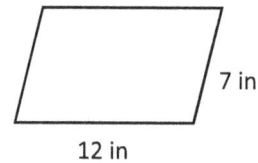

7 in
12 in

8) Square

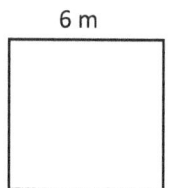

6 m

✎ **Find the area of each shape.**

9) Parallelogram

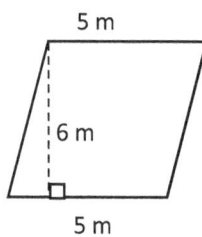

5 m
6 m
5 m

10) Rectangle

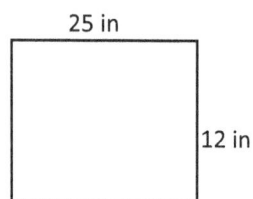

25 in
12 in

11) Rectangle

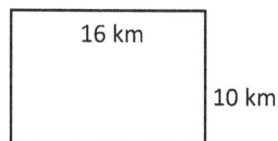

16 km
10 km

12) Square

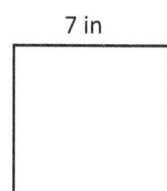

7 in

Trapezoids

✎ **Find the area of each trapezoid.**

1)

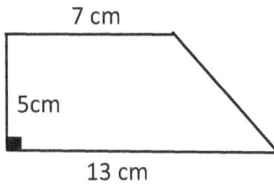

7 cm
5cm
13 cm

2)

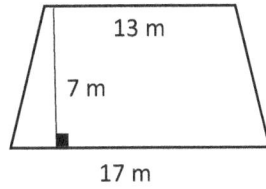

13 m
7 m
17 m

3)

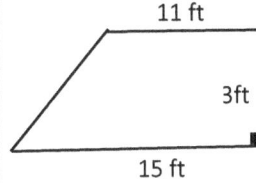

11 ft
3ft
15 ft

4)

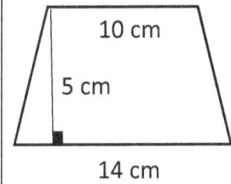

10 cm
5 cm
14 cm

5)

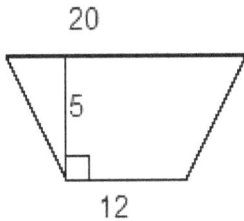

20
5
12

6)

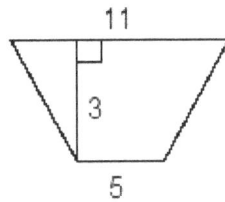

11
3
5

7)

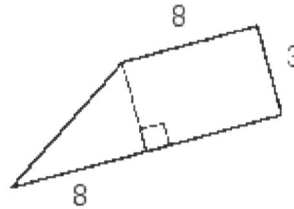

8
3
8

8)

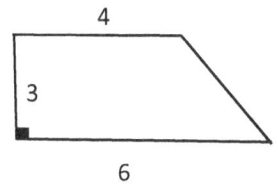

4
3
6

✎ **Calculate.**

1) A trapezoid has an area of 45 cm² and its height is 5 cm and one base is 5 cm. What is the other base length? _____

2) If a trapezoid has an area of 99 ft² and the lengths of the bases are 8 ft and 10 ft, find the height? _____

3) If a trapezoid has an area of 126 m² and its height is 14 m and one base is 6 m, find the other base length? _____

4) The area of a trapezoid is 440 ft² and its height is 22 ft. If one base of the trapezoid is 15 ft, what is the other base length?

_____

Circles

✎ **Find the area of each circle.** ($\pi = 3.14$)

1)	2)	3)	4)	5)	6)
2.5 in	5cm	9 ft	2 m	11 cm	10 miles

7)	8)	9)	10)	11)	12)
13 in	3 ft	4 m	12 cm	7 miles	8 ft

✎ **Complete the table below.** ($\pi = 3.14$)

Circle No.	Radius	Diameter	Circumference	Area
1	1 in	2 in	6.28 in	3.14 in^2
2		10 m		
3				28.26 ft^2
4			47.1 mi	
5		11 km		
6	7 cm			
7		12 ft		
8				314 m^2
9			56.52 in	
10	4.5 ft			

Cubes

✎ **Find the volume of each cube.**

1)	2)	3)	4)	5)	6)
	2 cm	6 ft	11 m	13 in	7 miles

7)	8	9)	10)	11)	12)
1.2 km	9 cm	2.1 ft	12 mm	0.2 in	0.1 km

✎ **Find the surface area of each cube.**

13)	14)	15)	16)	17)	18)
	7 m	5 ft	4.5 mm	1.1 km	11 cm

Rectangular Prism

✎ **Find the volume of each Rectangular Prism.**

1)

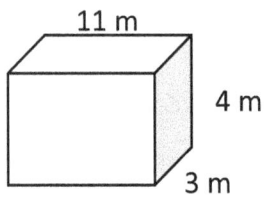

11 m

4 m

3 m

2)

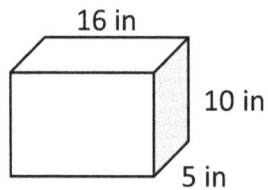

16 in

10 in

5 in

3)

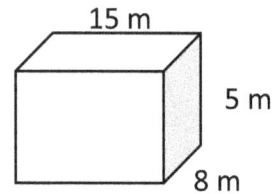

15 m

5 m

8 m

4)

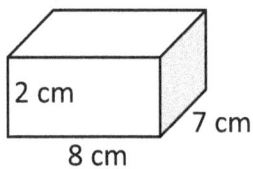

2 cm

7 cm

8 cm

5)

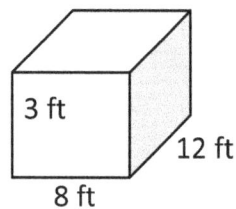

3 ft

12 ft

8 ft

6)

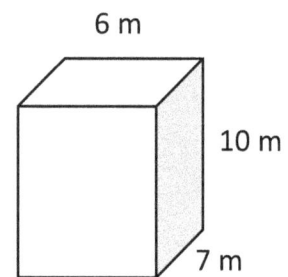

6 m

10 m

7 m

✎ **Find the surface area of each Rectangular Prism.**

7)

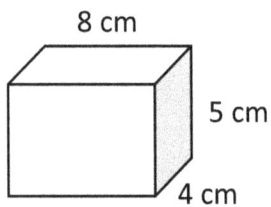

8 cm

5 cm

4 cm

8)

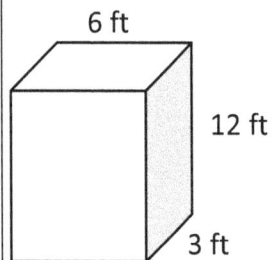

6 ft

12 ft

3 ft

9)

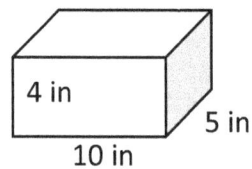

4 in

10 in

5 in

10)

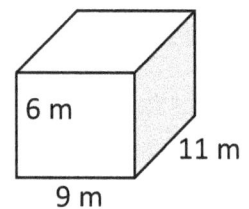

6 m

11 m

9 m

Cylinder

✎ **Find the volume of each Cylinder. Round your answer to the nearest tenth.** ($\pi = 3.14$)

1)

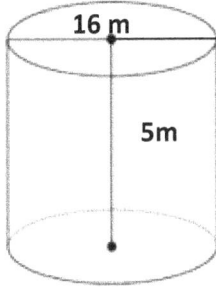

16 m

5m

2)

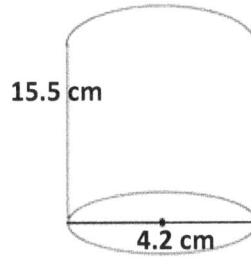

15.5 cm

4.2 cm

3)

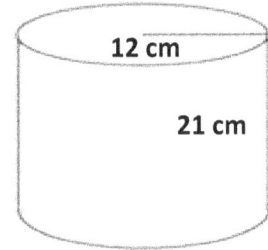

12 cm

21 cm

4)

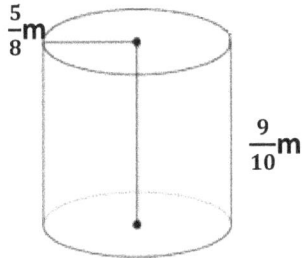

$\frac{5}{8}$m

$\frac{9}{10}$m

5)

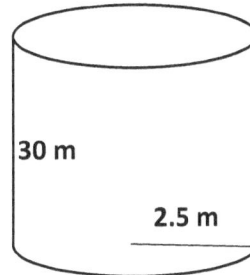

30 m

2.5 m

6)

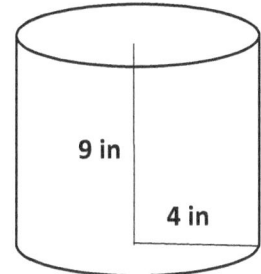

9 in

4 in

✎ **Find the surface area of each Cylinder.** ($\pi = 3.14$)

7)

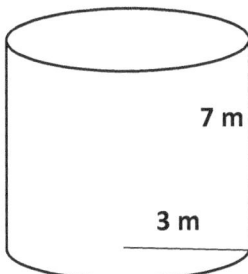

7 m

3 m

8)

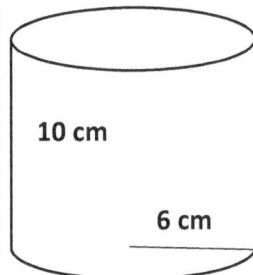

10 cm

6 cm

9)

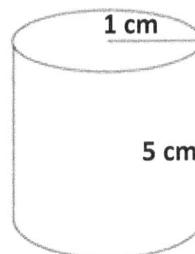

1 cm

5 cm

10)

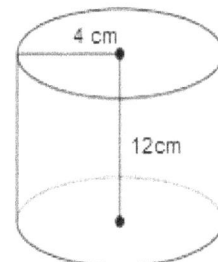

4 cm

12cm

Pyramids and Cone

✎ **Find the volume of each Pyramid and Cone.** ($\pi = 3.14$)

1)

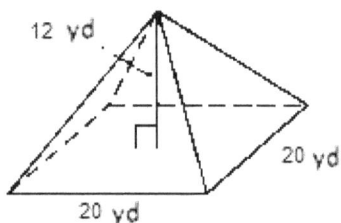

2)

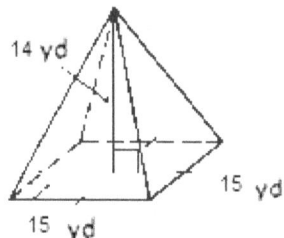

3)

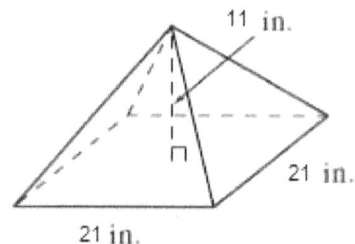

4)

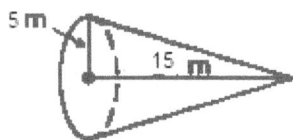

5)

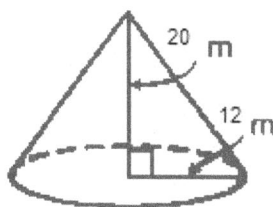

6)

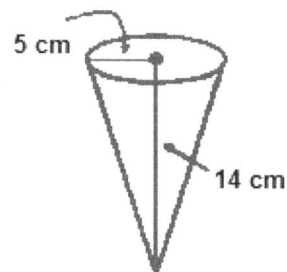

✎ **Find the surface area of each Pyramid and Cone.** ($\pi = 3.14$)

7)

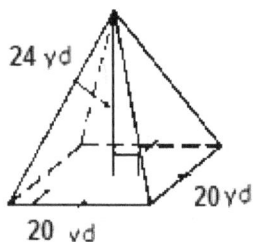

8)

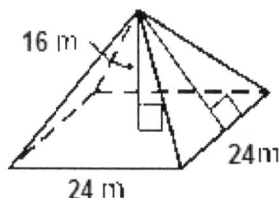

9)

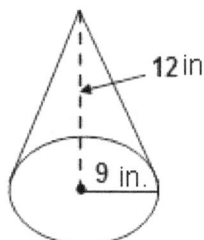

10)

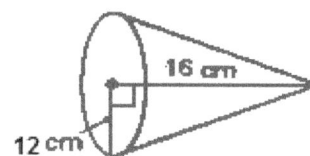

Answers of Worksheets

Angles

1) 16°	4) 34°	7) 90°	10) 75°
2) 96°	5) 70°	8) 75°	11) 70°
3) 59°	6) 52°	9) 33°	

Pythagorean Relationship

1) No	5) Yes	9) 13	13) 15
2) Yes	6) No	10) 20	14) 30
3) No	7) Yes	11) 17	15) 36
4) Yes	8) Yes	12) 10	16) 12

Triangles

1) 60°	5) 45°	9) 54 *square unites*
2) 48°	6) 40°	10) 120 *square unites*
3) 55°	7) 45°	11) 90 *square unites*
4) 52°	8) 67°	12) 180 *square unites*

Polygons

1) 52 *ft*	5) 54 *m*	9) 30 m^2
2) 26 *in*	6) 25 *cm*	10) 300 in^2
3) 112 *ft*	7) 38 *in*	11) 160 km^2
4) 20 *cm*	8) 24 *m*	12) 49 in^2

Trapezoids

1) 50 cm^2	4) 60 cm^2	7) 36
2) 105 m^2	5) 80	8) 15
3) 39 ft^2	6) 24	

Calculate

1) 13 *cm*	2) 11 *ft*	3) 12 *m*	4) 25 *ft*

Circles

1) 19.63 in^2	5) 379.94 cm^2	9) 12.56 m^2
2) 78.5 cm^2	6) 314 $miles^2$	10) 113.04 cm^2
3) 254.34 ft^2	7) 132.67 in^2	11) 38.47 $miles^2$
4) 12.56 m^2	8) 7.07 ft^2	12) 50.24 ft^2

Circle No.	Radius	Diameter	Circumference	Area
1	1 in	2 in	6.28 in	3.14 in^2
2	5 m	10 m	31.4 m	78.5 m^2
3	3 ft	6 ft	18.84 ft	28.26 ft^2
4	7.5 miles	15 mi	47.1 mi	176.63 mi^2
5	5.5 km	11 km	34.54 km	94.99 km^2
6	7 cm	14 cm	43.96 cm	153.86 cm^2
7	6 ft	12 ft	37.68 feet	113.04 ft^2
8	10 m	20 m	62.8 m	314 m^2
9	9 in	18 in	56.52 in	254.34 in^2
10	4.5 ft	9 ft	28.26 ft	63.585 ft^2

Cubes

1) 12
2) 8 cm^3
3) 216 ft^3
4) 1,331 m^3
5) 2,197 in^3
6) 343 $miles^3$
7) 1.728 km^3
8) 729 cm^3
9) 9.261 ft^3
10) 1,728 mm^3
11) 0.008 in^3
12) 0.001 km^3
13) 27
14) 294 m^2
15) 150 ft^2
16) 121.5 mm^2
17) 7.26 km^2
18) 726 cm^2

Rectangular Prism

1) 132 m^3
2) 800 in^3
3) 600 m^3
4) 112 cm^3
5) 288 ft^3
6) 420 m^3
7) 184 cm^2
8) 252 ft^2
9) 220 in^2
10) 438 m^2

Cylinder

1) 1,004.8 m^3
2) 214.6 cm^3
3) 9,495.4 cm^3
4) 1.1 m^3
5) 588.8 m^3
6) 452.2 in^3
7) 188.4 m^2
8) 602.9 cm^2
9) 37.7 cm^2
10) 401.9 m^2

Pyramids and Cone

1) 1,600 yd^3
2) 1,050 yd^3
3) 1,617 in^3
4) 392.5 m^3
5) 3,014.4 m^3
6) 366.33 cm^3
7) 1,440 yd^2
8) 1,536 m^2
9) 678.24 in^2
10) 1,205.76 cm^2

Chapter 11 :

Statistics and Probability

Topics that you will practice in this chapter:

- ✓ Mean and Median
- ✓ Mode and Range
- ✓ Histograms
- ✓ Stem–and–Leaf Plot
- ✓ Pie Graph
- ✓ Probability Problems
- ✓ Factorials
- ✓ Combinations and Permutation

Mathematics is no more computation than typing is literature.
– John Allen Paulos

Mean and Median

✎ **Find Mean and Median of the Given Data.**

1) 8, 7, 14, 4, 8

2) 14, 8, 25, 19, 16, 33, 11

3) 23, 18, 15, 12, 17

4) 34, 14, 10, 15, 6, 11

5) 10, 19, 6, 8, 32, 20, 17

6) 17, 26, 39, 69, 20, 6

7) 40, 38, 18, 11, 9, 2, 7, 32, 41

8) 24, 21, 31, 12, 33, 32, 22

9) 16, 14, 20, 41, 15, 20, 38, 4

10) 20, 20, 30, 18, 6, 28, 12, 46

11) 12, 7, 10, 11, 16, 22

12) 10, 29, 27, 12, 2, 15, 10, 3

✎ **Calculate.**

13) In a javelin throw competition, five athletics score 56, 34, 62, 23 and 19 meters. What are their Mean and Median? _____

14) Eva went to shop and bought 8 apples, 14 peaches, 6 bananas, 4 pineapples and 12 melons. What are the Mean and Median of her purchase? _____

15) Bob has 17 black pen, 19 red pen, 14 green pens, 20 blue pens and 5 boxes of yellow pens. If the Mean and Median are 19 respectively, what is the number of yellow pens in each box? _____

Mode and Range

✎ Find Mode and Rage of the Given Data.

1) 4, 3, 7, 3, 3, 4

   Mode: _____   Range: _____

2) 18, 18, 24, 26, 18, 8, 14, 22

   Mode: _____   Range: _____

3) 8, 8, 8, 16, 19, 22, 20, 9, 13

   Mode: _____   Range: _____

4) 24, 24, 14, 28, 20, 18, 20, 24

   Mode: _____   Range: _____

5) 6, 21, 27, 24, 27, 27

   Mode: _____   Range: _____

6) 21, 8, 8, 7, 8, 12, 10, 22, 18, 13

   Mode: _____   Range: _____

7) 7, 4, 4, 6, 13, 13, 13, 0, 2, 2

   Mode: _____   Range: _____

8) 5, 8, 5, 14, 12, 14, 3, 5, 18

   Mode: _____   Range: _____

9) 7, 7, 7, 12, 7, 3, 8, 16, 3, 17

   Mode: _____   Range: _____

10) 15, 15, 19, 16, 4, 16, 10, 15

   Mode: _____   Range: _____

11) 6, 6, 5, 6, 42, 13, 19, 2

   Mode: _____   Range: _____

12) 8, 8, 9, 8, 9, 4, 34, 22

   Mode: _____   Range: _____

✎ Calculate.

13) A stationery sold 12 pencils, 56 red pens, 24 blue pens, 20 notebooks, 12 erasers, 21 rulers and 11 color pencils. What are the Mode and Range for the stationery sells?

   Mode: _____   Range: _____

14) In an English test, eight students score 10, 15, 15, 18 18, 16, 15 and 15. What are their Mode and Range? _____

15) What is the range of the first 6 even numbers greater than 8?

   _____

Times Series

✑ **Use the following Graph to complete the table.**

Day	Distance (km)
1	
2	

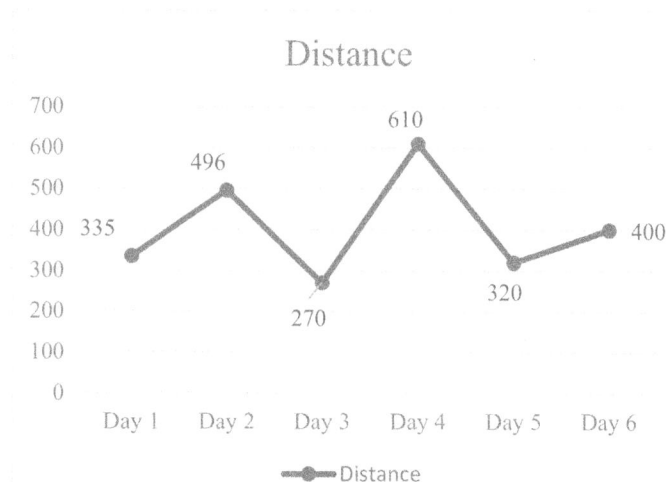

The following table shows the number of births in the US from 2007 to 2012 (in millions).

Year	Number of births (in millions)
2007	4.15
2008	3.70
2009	3.45
2010	3.20
2011	1.75
2012	2.98

Draw a Time Series for the table.

Stem–and–Leaf Plot

🖎 Make stem ad leaf plots for the given data.

1) 24, 26, 29, 20, 53, 27, 51, 55, 36, 21, 37, 30

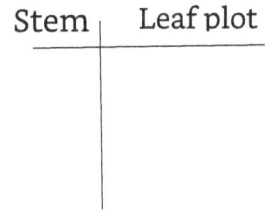

Stem	Leaf plot

2) 11, 59, 66, 14, 18, 19, 59, 65, 69, 61, 68, 65

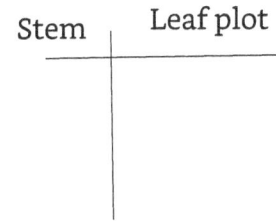

Stem	Leaf plot

3) 121, 55, 66, 54, 112, 128, 63, 125, 59, 123, 68, 119

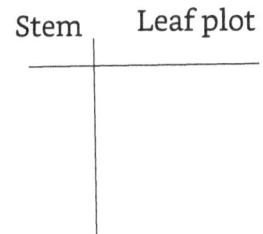

Stem	Leaf plot

4) 51, 32, 100, 56, 84, 36, 107, 56, 85, 39, 56, 106, 89

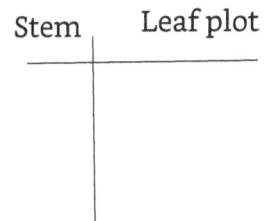

Stem	Leaf plot

5) 33, 89, 19, 87, 81, 16, 11, 30, 86, 35, 17, 35, 13

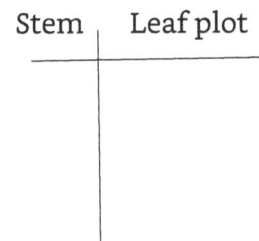

Stem	Leaf plot

6) 60, 92, 22, 25, 67, 93, 95, 62, 21, 64, 98, 29

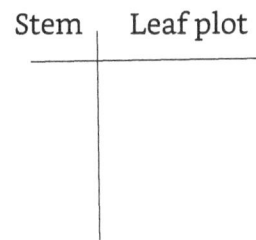

Stem	Leaf plot

Pie Graph

The circle graph below shows all Robert's expenses for last month. Robert spent $140 on his hobbies last month.

Answer following questions based on the Pie graph.

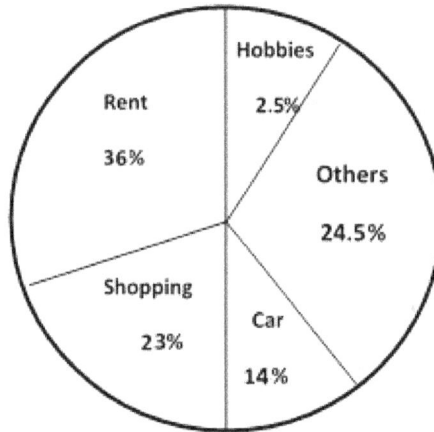

1) How much was Robert's total expenses last month? _____

2) How much did Robert spend on his car last month? _____

3) How much did Robert spend for shopping last month? _____

4) How much did Robert spend on his rent last month? _____

5) What fraction is Robert's expenses for his rent and car out of his total expenses last month? _____

Probability Problems

✎ **Calculate.**

1) A number is chosen at random from 1 to 10. Find the probability of selecting number 6 or smaller numbers. _____

2) Bag A contains 18 red marbles and 6 green marbles. Bag B contains 16 black marbles and 8 orange marbles. What is the probability of selecting a green marble at random from bag A? What is the probability of selecting a black marble at random from Bag B? _____

3) A number is chosen at random from 1 to 20. What is the probability of selecting multiples of 4? _____

4) A card is chosen from a well-shuffled deck of 52 cards. What is the probability that the card will be a queen? _____

5) A number is chosen at random from 1 to 15. What is the probability of selecting a multiple of 3 or 5? _____

A spinner numbered 1–8, is spun once. What is the probability of spinning ...?

6) an Odd number? _____   7) a multiple of 2? _____

8) a multiple of 5? _____   9) number 10? _____

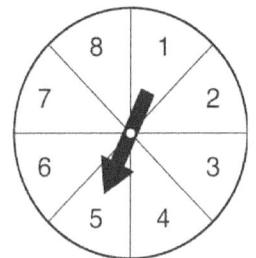

Factorials

✎ **Determine the value for each expression.**

1) $4! + 0! =$

2) $2! + 5! =$

3) $(2!)^2 =$

4) $5! - 3! =$

5) $6! - 3! + 10 =$

6) $3! \times 4 - 15 =$

7) $(2! + 3!)^2 =$

8) $(4! - 3!)^2 =$

9) $(3!\,0!)^2 - 10 =$

10) $\dfrac{10!}{8!} =$

11) $\dfrac{6!}{4!} =$

12) $\dfrac{6!}{5!} =$

13) $\dfrac{15!}{13!} =$

14) $\dfrac{n!}{(n-3)!} =$

15) $\dfrac{(n+2)!}{n!} =$

16) $\dfrac{(2+2!)^3}{2!} =$

17) $\dfrac{5(n+2)!}{(n+1)!} =$

18) $\dfrac{22!}{20!4!} =$

19) $\dfrac{13!}{11!3!} =$

20) $\dfrac{9\times210!}{3(7\times30)!} =$

21) $\dfrac{32!}{31!2!} =$

22) $\dfrac{11!12!}{10!13!} =$

23) $\dfrac{16!15!}{14!14!} =$

24) $\dfrac{(5\times3)!}{0!14!} =$

25) $\dfrac{4!(5n-2)!}{(5n)!} =$

26) $\dfrac{4n(4n+7)!}{(4n+8)!} =$

27) $\dfrac{(n-2)!(n+1)}{(n+2)!} =$

Combinations and Permutations

✍ Calculate the value of each.

1) $6! =$ ____

2) $2! \times 5! =$ ____

3) $3 \times 4! =$ ____

4) $5! + 3! =$ ____

5) $7! =$ ____

6) $4! =$ ____

7) $3! + 3! =$ ____

8) $7! - 5! =$ ____

✍ Find the answer for each word problems.

9) Susan is baking cookies. She uses sugar, butter, Vanilla, eggs and flour. How many different orders of ingredients can she try? _____

10) Albert is planning for his vacation. He wants to go to museum, watch a movie, go to the beach, play the game and play football. How many ways of ordering are there for him? _____

11) How many 4-digit numbers can be named using the digits 3, 4, 5, and 6 without repetition? _____

12) In how many ways can 5 boys be arranged in a straight line? _____

13) In how many ways can 6 athletes be arranged in a straight line? _____

14) A professor is going to arrange her 7 students in a straight line. In how many ways can she do this? _____

15) How many code symbols can be formed with the letters for the word GAMES? _____

16) In how many ways a team of 7 basketball players can choose a captain and co-captain? _____

Answers of Worksheets

Mean and Median

1) Mean: 8.2, Median: 8
2) Mean: 18, Median: 16
3) Mean: 17, Median: 17
4) Mean: 15, Median: 12.5
5) Mean: 16, Median: 17

6) Mean: 29.5, Median: 23
7) Mean: 22, Median: 18
8) Mean: 25, Median: 24
9) Mean: 21, Median: 18
10) Mean: 22.5, Median: 20

11) Mean: 13, Median: 11.5
12) Mean: 13.5, Median: 11
13) Mean: 38.8, Median: 34
14) Mean: 8.8, Median: 8
15) 5

Mode and Range

1) Mode: 3, Range: 4
2) Mode: 18, Range: 18
3) Mode: 8, Range: 14
4) Mode: 24, Range: 14
5) Mode: 27, Range: 21

6) Mode: 8, Range: 15
7) Mode: 13, Range: 13
8) Mode: 5, Range: 15
9) Mode: 7, Range: 14
10) Mode: 15, Range: 15

11) Mode: 6, Range: 40
12) Mode: 8, Range: 30
13) Mode: 12, Range: 45
14) Mode: 15, Range: 8
15) 10

Time series

Day	Distance (km)
1	335
2	496
3	270
4	610
5	320
6	400

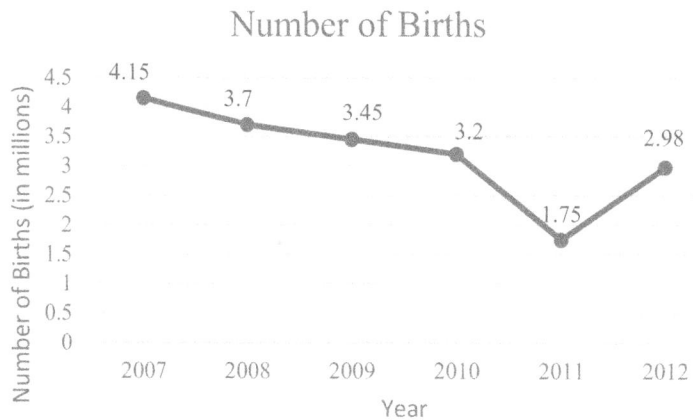

Number of Births

Number of Births (in millions)

4.15 3.7 3.45 3.2 1.75 2.98

2007 2008 2009 2010 2011 2012

Year

Stem–And–Leaf Plot

1)

Stem	leaf
2	0 1 4 6 7 9
3	0 6 7
5	1 3 5

2)

Stem	leaf
1	1 4 8 9
5	9 9
6	1 5 5 6 8 9

3)

Stem	leaf
5	4 5 9
6	3 6 8
11	2 9
12	1 3 5 8

4)

Stem	leaf
3	2 6 9
5	1 6 6 6
8	4 5 9
10	0 6 7

5)

Stem	leaf
1	1 3 6 7 9
3	0 3 5 5
8	1 6 7 9

6)

Stem	leaf
2	2 1 5 9
6	0 2 4 7
9	2 3 5 8

Pie Graph

1) $5,600

2) $784

3) $1,288

4) $2,016

5) $\frac{1}{2}$

Probability Problems

1) $\frac{3}{5}$

2) $\frac{1}{4}, \frac{2}{3}$

3) $\frac{1}{4}$

4) $\frac{1}{13}$

5) $\frac{7}{15}$

6) $\frac{1}{2}$

7) $\frac{1}{2}$

8) $\frac{1}{8}$

9) 0

Factorials

1) 25

2) 122

3) 4

4) 114

5) 724

6) 9

7) 64

8) 324

9) 26

10) 90

11) 30

12) 6

13) 210

14) $n(n-1)(n-2)$

15) $(n+1)(n+2)$

16) 32

17) $5(n+2)$

18) 19.25

19) 26

20) 3

21) 16

22) $\frac{11}{13}$

23) 3,600

24) 15

25) $\frac{24}{5n(5n-1)}$

26) $\frac{n}{(n+2)}$

27) $\frac{1}{n(n-1)(n+2)}$

Combinations and Permutations

1) 720

2) 240

3) 72

4) 126

5) 5,040

6) 24

7) 12

8) 4,920

9) 120

10) 120

11) 24

12) 120

13) 720

14) 5,040

15) 120

16) 42

Chapter 12 : SSAT Upper-Level Practice Tests

The SSAT, or Secondary School Admissions Test, is a standardized test to help determine admission to private elementary, middle and high schools.

There are currently three Levels of the SSAT:

- ✓ Lower Level (for students in 3rd and 4th grade)
- ✓ Middle Level (for students in 5th-7th grade)
- ✓ Upper-Level (for students in 8th-11th grade)

The SSAT Upper-Level test consists of three separate exams:

- ✓ Quantitative (Mathematics)
- ✓ Reading
- ✓ Writing

There are two quantitative sections on the test. The mathematics portions of the SSAT Upper-Level test each contains 25 questions. For each math section, students have 30 minutes to complete the test. The Math section of the test covers arithmetic, data analysis, geometry, algebra, and some basic statistics topics.

Students are not allowed to use calculator when taking a SSAT Upper-Level assessment.

In this section, there are two complete SSAT Upper-Level Mathematics Tests. Take these tests to see what score you'll be able to receive on a real SSAT Upper-Level test.

Time to Test

Time to refine your skill with a practice examination.

Take a practice SSAT Upper-Level Math Test to simulate the test day experience. After you've finished, score your test using the answer key.

Before You Start

- You'll need a pencil and a timer to take the test.

- After you've finished the test, review the answer key to see where you went wrong.

- Use the answer sheet provided to record your answers. (You can cut it out or photocopy it)

- You will receive 1 point for every correct answer, and you will lose $\frac{1}{4}$ point for each incorrect answer. There is no penalty for skipping a question.

    Calculators are NOT permitted for the SSAT Upper-Level Test

Good Luck!

SSAT Upper-Level Practice Test Answer Sheets
Remove (or photocopy) these answer sheets and use them to complete the practice tests.

SSAT Middle Level Practice Test

Mathematics Section 1

1	Ⓐ Ⓑ Ⓒ Ⓓ Ⓔ	16	Ⓐ Ⓑ Ⓒ Ⓓ Ⓔ
2	Ⓐ Ⓑ Ⓒ Ⓓ Ⓔ	17	Ⓐ Ⓑ Ⓒ Ⓓ Ⓔ
3	Ⓐ Ⓑ Ⓒ Ⓓ Ⓔ	18	Ⓐ Ⓑ Ⓒ Ⓓ Ⓔ
4	Ⓐ Ⓑ Ⓒ Ⓓ Ⓔ	19	Ⓐ Ⓑ Ⓒ Ⓓ Ⓔ
5	Ⓐ Ⓑ Ⓒ Ⓓ Ⓔ	20	Ⓐ Ⓑ Ⓒ Ⓓ Ⓔ
6	Ⓐ Ⓑ Ⓒ Ⓓ Ⓔ	21	Ⓐ Ⓑ Ⓒ Ⓓ Ⓔ
7	Ⓐ Ⓑ Ⓒ Ⓓ Ⓔ	22	Ⓐ Ⓑ Ⓒ Ⓓ Ⓔ
8	Ⓐ Ⓑ Ⓒ Ⓓ Ⓔ	23	Ⓐ Ⓑ Ⓒ Ⓓ Ⓔ
9	Ⓐ Ⓑ Ⓒ Ⓓ Ⓔ	24	Ⓐ Ⓑ Ⓒ Ⓓ Ⓔ
10	Ⓐ Ⓑ Ⓒ Ⓓ Ⓔ	25	Ⓐ Ⓑ Ⓒ Ⓓ Ⓔ
11	Ⓐ Ⓑ Ⓒ Ⓓ Ⓔ		
12	Ⓐ Ⓑ Ⓒ Ⓓ Ⓔ		
13	Ⓐ Ⓑ Ⓒ Ⓓ Ⓔ		
14	Ⓐ Ⓑ Ⓒ Ⓓ Ⓔ		
15	Ⓐ Ⓑ Ⓒ Ⓓ Ⓔ		

Mathematics Section 2

1	Ⓐ Ⓑ Ⓒ Ⓓ Ⓔ	16	Ⓐ Ⓑ Ⓒ Ⓓ Ⓔ
2	Ⓐ Ⓑ Ⓒ Ⓓ Ⓔ	17	Ⓐ Ⓑ Ⓒ Ⓓ Ⓔ
3	Ⓐ Ⓑ Ⓒ Ⓓ Ⓔ	18	Ⓐ Ⓑ Ⓒ Ⓓ Ⓔ
4	Ⓐ Ⓑ Ⓒ Ⓓ Ⓔ	19	Ⓐ Ⓑ Ⓒ Ⓓ Ⓔ
5	Ⓐ Ⓑ Ⓒ Ⓓ Ⓔ	20	Ⓐ Ⓑ Ⓒ Ⓓ Ⓔ
6	Ⓐ Ⓑ Ⓒ Ⓓ Ⓔ	21	Ⓐ Ⓑ Ⓒ Ⓓ Ⓔ
7	Ⓐ Ⓑ Ⓒ Ⓓ Ⓔ	22	Ⓐ Ⓑ Ⓒ Ⓓ Ⓔ
8	Ⓐ Ⓑ Ⓒ Ⓓ Ⓔ	23	Ⓐ Ⓑ Ⓒ Ⓓ Ⓔ
9	Ⓐ Ⓑ Ⓒ Ⓓ Ⓔ	24	Ⓐ Ⓑ Ⓒ Ⓓ Ⓔ
10	Ⓐ Ⓑ Ⓒ Ⓓ Ⓔ	25	Ⓐ Ⓑ Ⓒ Ⓓ Ⓔ
11	Ⓐ Ⓑ Ⓒ Ⓓ Ⓔ		
12	Ⓐ Ⓑ Ⓒ Ⓓ Ⓔ		
13	Ⓐ Ⓑ Ⓒ Ⓓ Ⓔ		
14	Ⓐ Ⓑ Ⓒ Ⓓ Ⓔ		
15	Ⓐ Ⓑ Ⓒ Ⓓ Ⓔ		

SSAT Upper-Level Practice Test 1

Quantitative 1

- ❖ **25 Questions.**
- ❖ **Total time for this test: 30 Minutes**.
- ❖ **Calculators are not allowed at the test**.

Administered

1) $58.265 \div 0.005$?

    A. 11,563   D. 11,653

    B. 10,356   E. 12,653

    C. 11.653

2) What is the value of the sum of the tens and thousandths in number

   7,953.21438?

    A. 8   D. 15

    B. 10   E. 13

    C. 12

3) If 140 % of a number is 84, then what is the 30 % of that number?

    A. 18   D. 25

    B. 30   E. 21

    C. 14

4) Jack earns $880 for his first 40 hours of work in a week and is then paid 1.5 times his regular hourly rate for any additional hours. This week, Jack needs $1,045 to pay his rent, bills and other expenses. How many hours must he work to make enough money in this week?

    A. 53   D. 35

    B. 42   E. 51

    C. 45

5) $\dfrac{2\frac{1}{8} + \frac{3}{4}}{1\frac{1}{5} - \frac{3}{10}}$ is approximately equal to?

A. 3.29

D. 3.19

B. 2.15

E. 2.18

C. 3.15

6) If $1 \le x < 6$, what is the minimum value of the following expression?

$$5x + 3$$

A. 18

D. 15

B. 8

E. 33

C. 5

7) A shaft rotates 360 times in 12 seconds. How many times does it rotate in 8 seconds?

A. 240

D. 5,760

B. 420

E. 1,440

C. 540

8) A school wants to give each of its 74 top students a football ball. If the balls are in boxes of 8, how many boxes of balls they need to purchase?

A. 8

D. 12

B. 11

E. 9

C. 10

9) If $\frac{36}{A} + 3 = 7$, then $32 + A = ?$

    A. 4   D. 41

    B. 23   E. 39

    C. 36

10) The average weight of 23 girls in a class is 31 kg and the average weight of 27 boys in the same class is 55 kg. What is the average weight of all the 50 students in that class?

    A. 29.7   D. 71

    B. 43.96   E. 33.85

    C. 54.95

11) There are five equal tanks of water. If $\frac{4}{7}$ of a tank contains 200 liters of water, what is the capacity of the five tanks of water together?

    A. 1,750   D. 350

    B. 1,250   E. 1,400

    C. 750

12) The sum of six different negative integers is -105. If the smallest of these integers is -20, what is the largest possible value of one of the other five integers?

    A. -35 D. -12

    B. -15 E. -27

    C. -10

13) Which of the following statements is correct, according to the graph below?

Number of Books Sold in a Bookstore

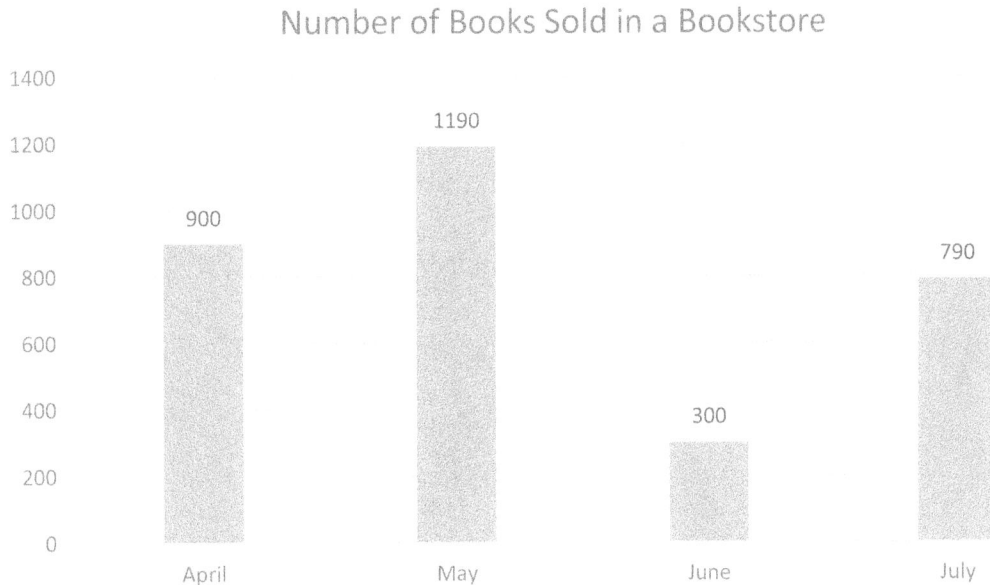

A. Number of books sold in June was greater than half the number of books sold in July.

B. Number of books sold in July was half the number of books sold in May.

C. Number of books sold in April was triple the number of books sold in June.

D. Number of books sold in July was equal to the number of books sold in April plus the number of books sold in June.

E. More books were sold in July than in April.

14) If $x \blacksquare y = \sqrt{x^2 + 3y}$, what is the value of $10 \blacksquare 7$?

A. $\sqrt{110}$

B. 11

C. 17

D. 9

E. 14

15) What is the answer of $8.4 \div 0.12$?

    A. $\frac{1}{70}$ D. 70

    B. $\frac{1}{7}$ E. 700

    C. 7

16) Four-kilograms apple and three-kilograms orange cost $54.4. If one-kilogram apple costs $3.1 how much does one-kilogram orange cost?

    A. $7   D. $14

    B. $21   E. $5.2

    C. $12.4

17) David's current age is 51 years, and Ava's current age is 7 years old. In how many years David's age will be 5 times Ava's age?

    A. 8   D. 12

    B. 6   E. 14

    C. 4

18) What is the value of x in the following figure?

    A. 135

    B. 125

    C. 45

    D. 117

    E. 108

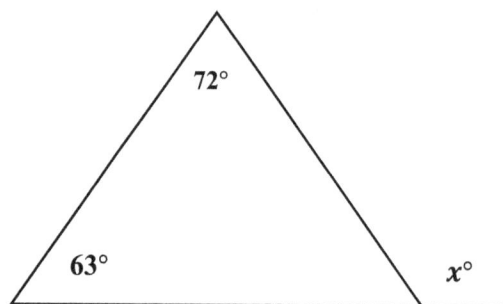

19) How many tiles of 5 cm^2 is needed to cover a floor of dimension 4 cm by 30 cm?

    A. 30

    B. 24

    C. 26

    D. 15

    E. 20

20) Michelle and Alec can finish a job together in 150 minutes. If Michelle can do the job by herself in 5 hours, how many minutes does it take Alec to finish the job?

    A. 210

    B. 150

    C. 600

    D. 300

    E. 340

21) A football team won exactly 80% of the games it played during last session. Which of the following could be the total number of games the team played last season?

    A. 16

    B. 50

    C. 48

    D. 63

    E. 34

22) The width of a box is one fifth of its length. The height of the box is one fourth of its width. If the length of the box is 60 cm, what is the volume of the box?

    A. 720 cm^3

    B. 36 cm^3

    C. 2,880 cm^3

    D. 2,160 cm^3

    E. 2,180 cm^3

23) What is the slope of a line that is perpendicular to the line $15x - 3y = 30$?

A. -3

D. $\frac{1}{5}$

B. $-\frac{1}{5}$

E. 15

C. -15

24) A cruise line ship left Port A and traveled 15 miles due west and then 20 miles due north. At this point, what is the shortest distance from the cruise to port A?

A. 35 miles

D. 25 miles

B. 32 miles

E. 30 miles

C. 50 miles

25) The Jackson Library is ordering some bookshelves. If x is the number of bookshelves the library wants to order, which each cost \$450 and there is a one-time delivery charge of \$840, which of the following represents the total cost, in dollar, per bookshelf?

A. $450x + 840$

D. $\frac{450x+840}{x}$

B. $450 + 840x$

E. $450x - 840$

C. $\frac{450x+840}{450}$

STOP
IF YOU FINISH BEFORE TIME IS CALLED, YOU MAY CHECK YOUR WORK ON THIS
SECTION ONLY. DO NOT TURN TO ANY OTHER SECTION IN THE TEST.

SSAT Upper-Level Practice Test 1

Quantitative 2

❖ **25 Questions.**

❖ **Total time for this test: 30 Minutes.**

❖ **Calculators are not allowed at the test.**

Administered *Month Year*

1) Which of the following expression is NOT equal to 8?

    A. $24 \times \frac{1}{3}$ D. $7 \times \frac{8}{7}$

    B. 40×0.2 E. $12 \times \frac{5}{12}$

    C. 5×1.6

2) When number 53,238 is divided by 325, the result is closest to?

    A. 16.8   D. 162

    B. 16.4   E. 165

    C. 164

3) If Jason's mark is k more than Alex, and Jason's mark is 23, which of the following can be Alex's mark?

    A. $23 + k$ D. 23k

    B. $k - 23$ E. $23 - k$

    C. $\frac{k}{23}$

4) To paint a wall with the area of $26m^2$, how many liters of paint do we need if each liter of paint is enough to paint a wall with dimension of $52 \text{ cm} \times 100 \text{ cm}$?

    A. 50   D. 140

    B. 40   E. 255

    C. 150

5) $150 - 14\frac{19}{42} = ?$

   A. $135\frac{5}{7}$ D. $135\frac{23}{42}$

   B. $136\frac{5}{21}$ E. $137\frac{23}{42}$

   C. $135\frac{19}{42}$

6) The price of a sofa is decreased by 30% to $455. What was its original price?

   A. $136.5   D. $360

   B. $560   E. $1,365

   C. $650

7) A driver rests three hour and 10 minutes for every 7 hours driving. How many minutes will he rest if he drives 21 hours?

   A. 9 hours

   B. 8 hours and 30 minutest

   C. 9 hours and 20 minutest

   D. 8 hours and 50 minutest

   E. 9 hours and 30 minutest

8) What is the missing term in the given sequence?

   4, 5, 7, 10, 14, 19, 25, ___, 40

   A. 27   D. 30

   B. 29   E. 32

   C. 31

Questions 9 and 10 are based on the following graph.

A library has 650 books that include Mathematics, Physics, Chemistry, English, and History.

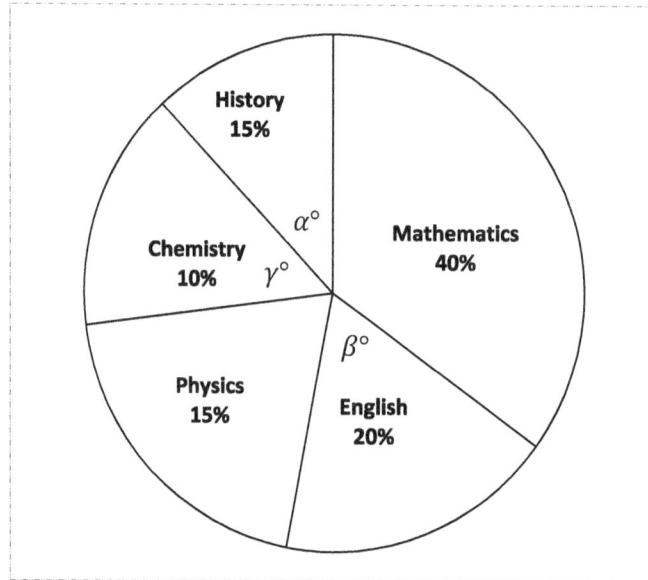

9) What is the product of the number of Mathematics and number of English books?

A. 38,300

D. 390

B. 33,800

E. 16,900

C. 33,600

10) What are the values of angle α and β respectively?

A. 46°, 36°

D. 54°, 72°

B. 54°, 144°

E. 52°, 74°

C. 36°, 72°

11) Find the perimeter of following shape.

A. 65

B. 68

C. 54

D. 64

E. 46

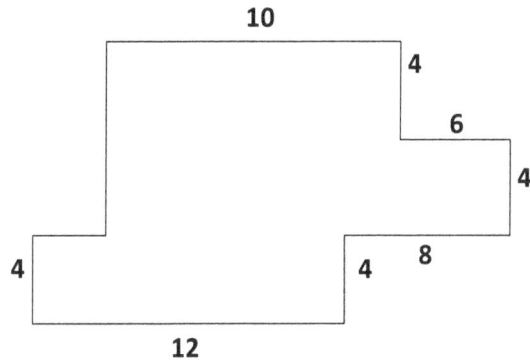

12) If $6y + 8 < 44$, then y could be equal to?

A. 12   D. 5

B. 9.5   E. 9

C. 6.5

13) The capacity of a red box is 60% bigger than the capacity of a blue box. If the red box can hold 77 equal sized books, how many of the same books can the blue box hold?

A. 24

B. 36

C. 28

D. 48

E. 20

14) If the area of the following rectangular ABCD is 200, and E is the midpoint of AB, what is the area of the shaded part?

A. 50

B. 20

C. 100

D. 200

E. 150

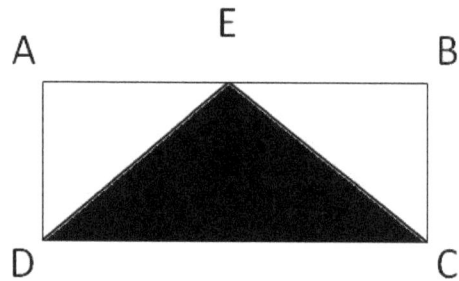

15) There are 12 marbles in the bag A and 24 marbles in the bag B. If the sum of the marbles in both bags will be shared equally between two children, how many marbles bag A has less than the marbles that each child will receive?

A. 6   D. 5

B. 4   E. 8

C. 3

16) There are 51.8 liters of gas in a car fuel tank. In the first week and second week of April, the car uses 8.4 and 27.6 liters of gas respectively. If the car was park in the third week of April and 16.54 liters of gas will be added to the fuel tank, how many liters of gas are in the fuel tank of the car?

A. 43.25   D. 347.25

B. 32.34   E. 42.35

C. 28

17) If the perimeter of the following figure be 24, what is the value of x?

A. 2

B. 3

C. 5

D. 4

E. 1

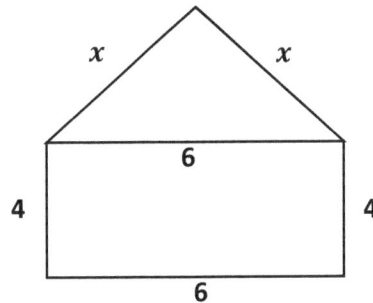

18) If $7x + y = 44$ and $2x - 5z = 16$, what is the value of x?

A. 12

C. 8

B. 14

D. 16

E. it cannot be determined from the information given

19) If a × b is divisible by 3, which of the following expression must also be divisible by 3?

A. $5a + b$

D. $\frac{2a}{3b}$

B. $2a + b$

E. $6 \times a \times b$

C. $a + 6b$

20) What is the average of circumference of figure A and area of figure B? ($\pi = 3$)

A. 30

B. 45

C. 62

D. 40

E. 50

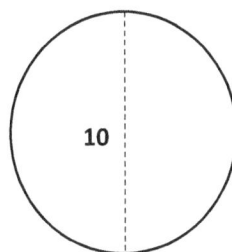

Figure A

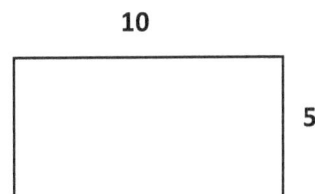

Figure B

21) In the following figure, point Q lies on the line n, what is the value of y if $x = 20$?

A. 15

B. 45

C. 30

D. 10

E. 40

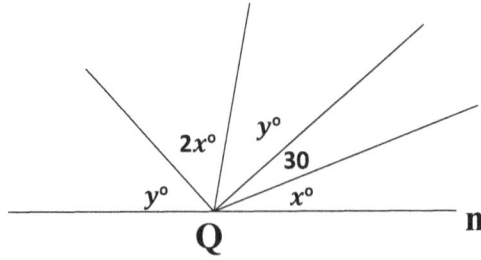

22) Which of the following could be the value of x if $\frac{3}{8} + x > 1$?

A. $\frac{1}{6}$

D. $\frac{1}{9}$

B. $\frac{1}{5}$

E. $\frac{4}{5}$

C. $\frac{3}{7}$

23) $565 \div 3 = ?$

A. $\frac{500}{3} \times \frac{60}{3} \times \frac{5}{3}$

D. $\frac{500}{3} \div \frac{60}{3} \div \frac{5}{3}$

B. $500 + \frac{60}{3} + \frac{5}{3}$

E. $\frac{5}{3} + \frac{6}{3} + \frac{5}{3}$

C. $\frac{500}{3} + \frac{60}{3} + \frac{5}{3}$

24) If a gas tank can hold 45 gallons, how many gallons does it contain when it is $\frac{5}{9}$ full?

A. 25

D. 10

B. 20

E. 30

C. 15

25) A number is chosen at random from 1 to 10. Find the probability of not selecting

a composite number (A composite number is a number that is divisible by itself,

1 and at least one other whole number).

A. $\frac{3}{15}$

B. $\frac{1}{5}$

C. $\frac{2}{5}$

D. 1

E. $\frac{1}{3}$

STOP

IF YOU FINISH BEFORE TIME IS CALLED, YOU MAY CHECK YOUR WORK ON THIS SECTION ONLY. DO NOT TURN TO ANY OTHER SECTION IN THE TEST.

SSAT Upper-Level Practice Test 2

Quantitative 1

- ❖ **25 Questions.**
- ❖ **Total time for this test: 30 Minutes**.
- ❖ **Calculators are not allowed at the test**.

.

Administered *Month Year*

1) What is the value of the "7" in number 358.792?

    A. 7 ones   D. 7 tens

    B. 7 tenths   E. 7 thousandths

    C. 7 hundredths

2) $0.06 \times 25.00 =$?

    A. 2.5   D. 155

    B. 15.50   E. 0.155

    C. 1.5

3) Two ninth of 45 is equal to $\frac{5}{7}$ of what number?

    A. 24   D. 28

    B. 52   E. 7

    C. 14

4) If $x - 49 = -45$, then $x \times 3 =$?

    A. 16   D. 20

    B. 6   E. 12

    C. 14

5) Mia plans to buy a bracelet for every one of her 53 friends for their party. There are six bracelets in each pack. How many packs must she buy?

    A. 4   D. 16

    B. 5   E. 12

    C. 9

6) If Logan ran 4.4 miles in one fifth an hour, his average speed was?

   A. 0.22 miles per hour

   B. 10.5 miles per hour

   C. 4.2 miles per hour

   D. 22 miles per hour

   E. 11 miles per hour

7) A pizza maker has x lb. of flour to make pizzas. After he has used 68 lb. of flour, how much flour is left? The expression that correctly represents the quantity of flour left is:

   A. $68 + x$ D. $x - 68$

   B. $\frac{68}{x}$ E. $68x$

   C. $68 - x$

8) The distance between cities A and B is approximately 6,758 miles. If Alice drives an average of 69 miles per hour, how many hours will it take Alice to drive from city A to city B?

   A. Approximately 98 hours

   B. Approximately 95 hours

   C. Approximately 89 hours

   D. Approximately 96 hours

   E. Approximately 87 hours

9) Given the diagram, what is the perimeter of the quadrilateral?

A. 60

B. 47

C. 96

D. 264

E. 1,450

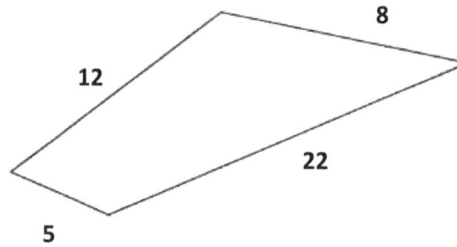

10) In a classroom of 80 students,56 are female. What percentage of the class is male?

A. 25%   D. 20%

B. 30%   E. 29%

C. 34%

11) An employee's rating on performance appraisals for the last four quarters were 67, 63, 61 and 69. If the required yearly average to qualify for the promotion is 68, what rating should the fifth quarter be?

A. 70   D. 80

B. 93   E. 69

C. 50

12) A steak dinner at a restaurant costs $4.65. If a man buys a steak dinner for himself and 3 friends, what will the total cost be?

A. $18.6   D. $15.50

B. $13.8   E. 12

C. $24

13) A cruise line ship left Port A and traveled 15 miles due west and then 36 miles due north. At this point, what is the shortest distance from the cruise to port A?

A. 39 miles

D. 46 miles

B. 114 miles

E. 48 miles

C. 41 miles

Questions 14 to 16 are based on the following data.

The result of a research shows the number of men and women in four cities of a country.

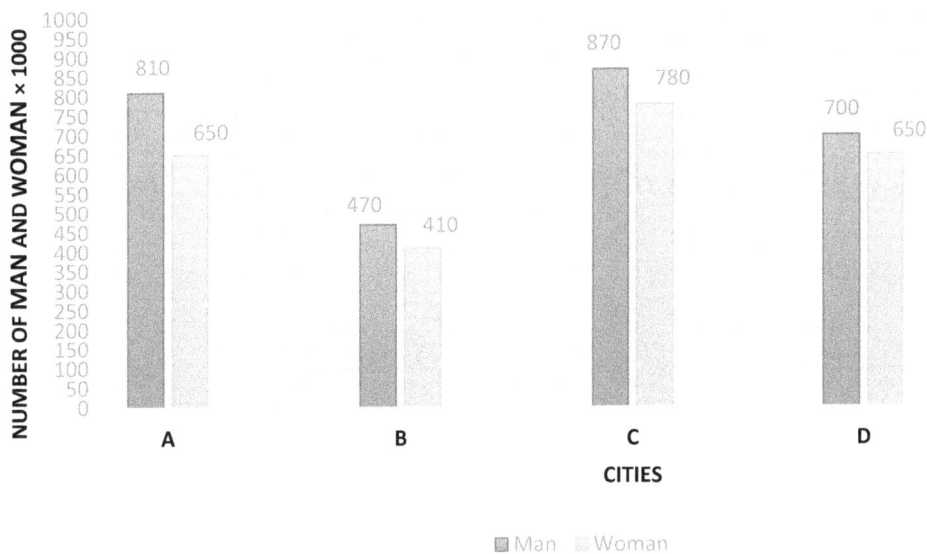

14) What's the ratio of percentage of men in city C to percentage of women in city B?

A. 0.55

D. 1.05

B. 0.105

E. 1.015

C. 0.15

15) What's the minimum ratio of men to women in the four cities?

    A. 0.68

    B. 0.56

    C. 0.58

    D. 0.80

    E. 1.02

16) How many women should be added to city C until the ratio of women to men will be 1.5?

    A. 256

    B. 255

    C. 525

    D. 656

    E. 555

17) Last week 15,000 fans attended a football match. This week triple as many bought tickets, but one ninth of them cancelled their tickets. How many are attending this week?

    A. 5,000

    B. 45,000

    C. 40,000

    D. 50,000

    E. 55,000

18) What is the slope of the line that is perpendicular to the line with equation $7x + y = 32$?

    A. $\frac{1}{7}$

    B. $-\frac{1}{7}$

    C. $\frac{32}{7}$

    D. 7

    E. -7

19) In the following figure, MN is 24 cm. How long is ON?

A. 12 cm

B. 14 cm

C. 18 cm

D. 10 cm

E. 20 cm

M   O   N

$5x$ $7x$

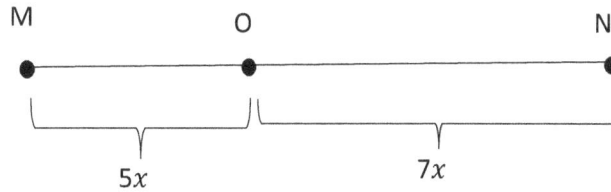

20) Solve the following equation for y?

$$\frac{x}{3+8} = \frac{y}{22-14}$$

A. $\frac{8}{11}x$

B. $\frac{11}{3}x$

C. $11x$

D. $14x$

E. $\frac{14}{3}x$

21) $\sqrt[6]{x^{55}} = ?$

A. $x^9\sqrt[6]{x}$

B. $9\sqrt[6]{x}$

C. $6x^{9.6}$

D. x^{52}

E. $x^6\sqrt{x}$

22) John traveled 240 km in 8 hours and Alice traveled 420 km in 6 hours. What is

the ratio of the average speed of John to average speed of Alice?

A. 5: 3

B. 3: 5

C. 3: 7

D. 2: 7

E. 7: 2

23) What is the difference of smallest 4–digit number and biggest 4–digit number?

    A. 6,966   D. 8,999

    B. 7,889   E. 9,999

    C. 8,888

24) Rectangle A has a length of 8 cm and a width of 4 cm, and rectangle B has a length of 3 cm and a width of 7 cm, what is the percent of ratio of the perimeter of rectangle B to rectangle A?

    A. 93%   D. 83%

    B. 95%   E. 103.1%

    C. 52%

25) In 1999, the average worker's income increased $5,000 per year starting from $28,000 annual salary. Which equation represents income greater than average? (I = income, x = number of years after 1999)

    A. $I > 5{,}000\,x + 28{,}000$

    B. $I > -5{,}000\,x + 28{,}000$

    C. $I < 5{,}000\,x - 28{,}000$

    D. $I < 28{,}000\,x - 5{,}000$

    E. $I < 28{,}000\,x + 5{,}000$

STOP

YOU FINISH BEFORE TIME IS CALLED; YOU MAY CHECK YOUR WORK ON THIS SECTION ONLY. DO NOT TURN TO ANY OTHER SECTION IN THE TEST.

SSAT Upper-Level Practice Test 2

Quantitative 2

❖ **25 Questions.**

❖ **Total time for this test: 30 Minutes**.

❖ **Calculators are not allowed at the test.**

Administered Month Year

1) $0.46 \times 11.7 = ?$

   A. 5.382   D. 5.328

   B. 5.822   E. 5.628

   C. 4.56

2) $3\frac{1}{5} \times 4\frac{1}{2} = ?$

   A. $14\frac{1}{5}$ D. $14\frac{5}{10}$

   B. $7\frac{5}{10}$ E. $12\frac{1}{2}$

   C. $14\frac{2}{10}$

3) If $\frac{2}{5}$ of a number equal to 52 then $\frac{7}{10}$ of the same number is:

   A. 80   D. 90

   B. 73   E. 101

   C. 91

4) Which of the following is a whole number ?

   A. $\frac{5}{6} \times \frac{2}{7}$ D. $5.4 + 3$

   B. $\frac{1}{4} + \frac{1}{5}$ E. $4.5 + \frac{7}{2}$

   C. $\frac{115}{15}$

5) We can put 18 colored pencils in each box, and we have 360 colored pencils.

How many boxes do we need?

A. 48

D. 15

B. 60

E. 20

C. 40

6) Sophia purchased a sofa for $360.10. The sofa is regularly priced at $554.

What was the percent discount Sophia received on the sofa?

A. 65%

D. 35%

B. 50%

E. 30%

C. 1.35%

7) How long does a 500–miles trip take moving at 40 miles per hour (mph)?

A. 13 hours

D. 13 hours and 20 minutes

B. 11 hours and 30 minutes

E. 10 hours and 30 minutes

C. 12 hours and 30 minutes

8) A swimming pool holds 5,600 cubic feet of water. The swimming pool is 14 feet

long and 5 feet wide. How deep is the swimming pool?

A. 80 feet

D. 130 feet

B. 60 feet

E. 160 feet

C. 20 feet

9) A bank is offering 2.45% simple interest on a savings account. If you deposit $27,000, how much interest will you earn in six years?

A. $1,069

B. $2,669

C. $6,039

D. $3,969

E. $769

10) 7 cubed is the same as:

A. 49

B. 14

C. $7 \times 7 \times 7 \times 7$

D. 343

E. 248

11) In the figure below, line A is parallel to line B. What is the value of angle x?

A. 45 degree

B. 55 degree

C. 125 degree

D. 42 degree

E. 122 degree

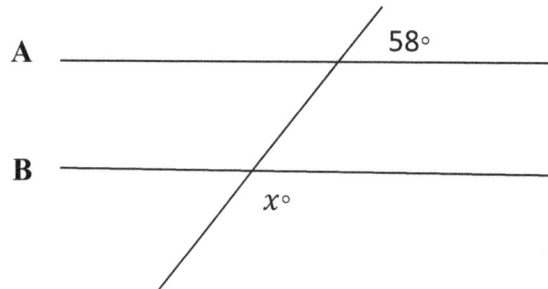

12) A construction company is building a wall. The company can build 20 cm of the wall per minute. After 30 minutes construction, $\frac{4}{5}$ of the wall is completed. How high is the wall?

A. 15 m

B. 75 m

C. 7.5 m

D. 30 m

E. 60 m

13) If $y = 5ab + 3b^3$, what is y when $a = 4$ and $b = 3$?

    A. 141   D. 81

    B. 91   E. 141

    C. 60

14) When a number is subtracted from 56 and the difference is divided by that number, the result is 7. What is the value of the number?

    A. 8   D. 22

    B. 11   E. 5

    C. 7

15) If car A drives 350 miles in 14 hours and car B drives the same distance in 7 hours, how many miles per hour does car B drive faster than car A?

    A. 10   D. 20

    B. 15   E. 25

    C. 30

16) The ratio of boys to girls in a school is 7: 3. If there are 540 students in a school, how many boys are in the school?

    A. 320   D. 400

    B. 350   E. 420

    C. 390

17) $\frac{(9+5)^2}{7} + 6 = ?$

    A. 34   D. 36

    B. 24   E. $\frac{13}{7}$

    C. 52

18) The area of a circle is 25π. What is the circumference of the circle?

    A. 5π D. 20π

    B. 12π E. 45π

    C. 10π

19) If 60% of x equal to 40% of 15, then what is the value of $(x + 5)^2$?

    A. 2.55   D. 225

    B. 15   E. 2,520

    C. 25.05

20) What is the greatest common factor of 24 and 72?

    A. 36   D. 6

    B. 12   E. 54

    C. 24

21) The length of a rectangle is $\frac{5}{9}$ times its width. If the width is 36, what is the perimeter of this rectangle?

    A. 120   D. 102

    B. 121   E. 72

    C. 112

22) Find $\frac{1}{4}$ of $\frac{4}{7}$ of 147?

    A. 22

    B. 21

    C. 36

    D. 12

    E. 48

23) If the interior angles of a quadrilateral are in the ratio 3:4:5:6, what is the measure of the smallest angle?

    A. 30°

    B. 60°

    C. 20°

    D. 18°

    E. 80°

24) A company pays its employee $8,000 plus 5% of all sales profit. If x is the number of all sales profit, which of the following represents the employee's revenue?

    A. $0.05\,x$

    B. $0.95\,x - 8,000$

    C. $0.05\,x + 8,000$

    D. $0.95\,x + 8,000$

    E. $5x + 8,000$

25) Which of the following shows the numbers in increasing order?

    A. $\frac{3}{11}, \frac{4}{7}, \frac{9}{13}, \frac{4}{5}$

    B. $\frac{3}{11}, \frac{4}{5}, \frac{9}{13}, \frac{4}{7}$

    C. $\frac{9}{13}, \frac{4}{7}, \frac{4}{5}, \frac{3}{11}$

    D. $\frac{3}{11}, \frac{9}{13}, \frac{4}{5}, \frac{4}{7}$

    E. None of the above

STOP

IF YOU FINISH BEFORE TIME IS CALLED, YOU MAY CHECK YOUR WORK ON THIS SECTION ONLY. DO NOT TURN TO ANY OTHER SECTION IN THE TEST.

Chapter 13 : Answers and Explanations

SSAT Upper-Level Practice Tests

Answer Key

❋ Now, it's time to review your results to see where you went wrong and what areas you need to improve!

SSAT Upper-Level Practice Test 1

Quantitative 1				Quantitative 2			
1	D	16	D	1	E	16	B
2	C	17	C	2	C	17	C
3	A	18	A	3	E	18	E
4	C	19	B	4	A	19	E
5	D	20	D	5	D	20	D
6	B	21	B	6	C	21	B
7	A	22	D	7	E	22	E
8	C	23	B	8	E	23	C
9	D	24	D	9	B	24	A
10	B	25	D	10	D	25	C
11	A			11	D		
12	B			12	D		
13	C			13	D		
14	B			14	C		
15	D			15	A		

Answers and Explanations

SSAT Upper-Level Practice Tests

SSAT Upper-Level Practice Test 2

Quantitative 1

1	B	16	C
2	C	17	C
3	C	18	A
4	E	19	B
5	C	20	A
6	D	21	A
7	D	22	C
8	A	23	D
9	B	24	D
10	B	25	A
11	D		
12	A		
13	A		
14	E		
15	D		

Quantitative 2

1	A	16	E
2	C	17	A
3	C	18	C
4	E	19	D
5	E	20	C
6	D	21	C
7	C	22	B
8	A	23	B
9	D	24	C
10	D	25	A
11	E		
12	C		
13	E		
14	C		
15	E		

Score Your Test

SSAT scores are broken down by its three sections: Verbal, Quantitative (or Math), and Reading. A sum of the three sections is also reported.

For the Upper-Level SSAT, the score range is 500-800, the lowest possible score a student can earn is 500 and the highest score is 800 for each section. A student receives 1 point for every correct answer and loses $\frac{1}{4}$ point for each incorrect answer.

No points are lost by skipping a question.

The total scaled score for an Upper-Level SSAT is the sum of the scores for the quantitative, verbal, and reading sections. A student will also receive a percentile score of between 1-99% that compares that student's test scores with those of other test takers of same grade and gender from the past 3 years.

Use the following table to convert SSAT Upper-level raw score to scaled score.

SSAT Upper-Level Math Scaled Scores	
Raw Scores	**Scaled Score**
50	800
45	790
40	765
35	740
30	715
25	694
20	668
15	642
10	605
5	575
0	544
-5	512
- 10 and lower	500

Answers and Explanations

SSAT - Upper-Level

Practice Tests 1: Quantitative 1

1) Answer: D.

$$58.265 \div 0.005 = \frac{\frac{58,265}{1,000}}{\frac{5}{1,000}} = \frac{58,265}{5} = 11,653$$

2) Answer: C.

The digit in tens place is 5.

The digit in the thousandths place is 7.

Therefore; $5 + 7 = 12$

3) Answer: A.

First, find the number.

Let x be the number. Write the equation and solve for x.

140 % of a number is 84, then: $1.4 \times x = 84 \rightarrow x = 84 \div 1.4 = 60$

30 % of 60 is: $0.3 \times 60 = 18$

4) Answer: C.

The amount of money that Jack earns for one hour: $\frac{\$880}{40} = \$22$

Number of additional hours that he needs to work in order to make enough money

is: $\frac{\$1,045-\$880}{1.5\times\$22} = 5$; Number of total hours is: $40 + 5 = 45$

5) Answer: D.

$$\frac{2\frac{1}{8}+\frac{3}{4}}{1\frac{1}{5}-\frac{3}{10}} = \frac{\frac{17}{8}+\frac{3}{4}}{\frac{6}{5}-\frac{3}{10}} = \frac{\frac{17+6}{8}}{\frac{12-3}{10}} = \frac{\frac{23}{8}}{\frac{9}{10}} = \frac{23\times10}{8\times9} = \frac{230}{72} \cong 3.19$$

6) Answer: B.

$1 \leq x < 6 \rightarrow$ Multiply all sides of the inequality by 5.

Then: $5 \times 1 \leq 5 \times x < 5 \times 6 \rightarrow 5 \leq 5x < 30$

Add 3 to all sides. $5 + 3 \leq 5x + 3 < 30 + 3 \rightarrow 8 \leq 5x + 3 < 33$

Minimum value of $5x + 3$ is 8

7) Answer: A.

Number of rotates in 8 second equals to: $\frac{360 \times 8}{12} = 240$

8) Answer: C.

Number of packs equal to: $\frac{74}{8} \cong 9.25$; Therefore, the school must purchase 10 packs.

9) Answer: D.

$\frac{36}{A} + 3 = 7 \rightarrow \frac{36}{A} = 7 - 3 = 4 \rightarrow 36 = 4A \rightarrow A = \frac{36}{4} = 9$

$32 + A = 32 + 9 = 41$

10) Answer: B.

$\text{Average} = \frac{\text{sum of terms}}{\text{number of terms}}$

The sum of the weight of all girls is: $23 \times 31 = 713$ kg

The sum of the weight of all boys is: $27 \times 55 = 1,485$ kg

The sum of the weight of all students is: $713 + 1,485 = 2,198$ kg

The average weight of the 50 students: $\frac{2,198}{50} = 43.96$

11) Answer: A.

Let x be the capacity of one tank. Then, $\frac{4}{7}x = 200 \rightarrow x = \frac{200 \times 7}{4} = 350$ Liters

The amount of water in five tanks is equal to: $5 \times 350 = 1,750$ Liters.

12) Answer: B.

The smallest number is -20. To find the largest possible value of one of the other five integers, we need to choose the smallest possible integers for five of them. Let x be the largest number. Then: $-105 = (-20) + (-19) + (-18) + (-17) + (-16) + x$

$\rightarrow -105 = -90 + x \rightarrow x = -105 + 90 = -15$

13) Answer: C.

Let's review the choices provided:

A. number of books sold in June is: 300

Half the number of books sold in July is: $\frac{790}{2} = 395 \rightarrow 300 < 395$

B. number of books sold in May is: 1,190

Half the number of books sold in May is: $\frac{1,190}{2} = 595 \rightarrow 595 \neq 790$

C. Number of books sold in April is: 900

Number of books sold in June is: $300 \rightarrow \frac{900}{300} = 3$

D. $900 + 300 = 1,200 > 790$

E. $790 < 900$; Only choice C is correct.

14) Answer: **B.**

$10 \blacksquare 7 = \sqrt{10^2 + 3(7)} = \sqrt{100 + 21} = \sqrt{121} = 11$

15) Answer: **D.**

$8.4 \div 0.12 = \frac{8.4}{0.12} = \frac{\frac{84}{10}}{\frac{12}{100}} = \frac{84 \times 100}{12 \times 10} = \frac{84}{12} \times \frac{100}{10} = 7 \times 10 = 70$

16) Answer: **D.**

Let x be the cost of one-kilogram orange, then: $3x + (4 \times 3.1) = 54.4$

$\rightarrow 3x + 12.4 = 54.4 \rightarrow 3x = 54.4 - 12.4 \rightarrow 3x = 42 \rightarrow x = \frac{42}{3} = \$14$

17) Answer: **C.**

Let's review the choices provided.

A. 8. In 8 years, David will be 59 and Ava will be 15. 59 is not 5 times 15.

B. 6. In 6 years, David will be 57 and Ava will be 13. 57 is not 5 times 13.

C. 4. In 4 years, David will be 55 and Ava will be 11. 55 is 5 times 11!

D. 12. In 12 years, David will be 63 and Ava will be 19. 63 is not 5 times 19.

E. 14. In 14 years, David will be 65 and Ava will be 21. 65 is not 5 times 21.

18) Answer: **A.**

All angles in a triangle sum up to 180 degrees. Then: $x = 63 + 72 = 135$

19) Answer: **B.**

The area of the floor is: $4cm \times 30cm = 120cm^2$

The number of tiles needed $= 120 \div 5 = 24$

20) Answer: **D.**

Let b be the amount of time Alec can do the job, then,

$$\frac{1}{a} + \frac{1}{b} = \frac{1}{150} \rightarrow \frac{1}{300} + \frac{1}{b} = \frac{1}{150} \rightarrow \frac{1}{b} = \frac{1}{150} - \frac{1}{300} = \frac{2-1}{300} = \frac{1}{300}$$

Then: b = 300 minutes

21) **Answer: B.**

Choices A, C and D are incorrect because 80% of each of the numbers is non-whole number.

A. 16,   80% of 16 = 0.80 × 16 = 12.8

B. 50,   80% of 50 = 0.80 × 50 = 40

C. 48,   80% of 48 = 0.80 × 48 = 38.4

D. 63,   80% of 63 = 0.80 × 63 = 50.4

E. 34,   80% of 34 = 0.80 × 34 = 27.2

22) **Answer: D.**

If the length of the box is 60, then the width of the box is one fifth of it, 12, and the height of the box is 3 (one fourth of the width). The volume of the box is:

V = (length)(width)(height) = (60) (12) (3) = 2,160 cm^3

23) **Answer: B.**

The equation of a line in slope intercept form is: y = mx + b. Solve for y.

15x − 3y = 30 ⇒ −3y = 30 − 15x ⇒ y = (30 − 15x) ÷ (−3) ⇒ y = 5x − 10

The slope is 5. The slope of the line perpendicular to this line is:

$m_1 \times m_2 = -1 \Rightarrow 5 \times m_2 = -1 \Rightarrow m_2 = -\frac{1}{5}$

24) **Answer: D.**

Use the information provided in the question to draw the shape.

Use Pythagorean Theorem: a^2 + b^2 = c^2

15^2 + 20^2 = c^2 ⇒ 225 + 400 = c^2 ⇒ 625 = c^2 ⇒ c = 25

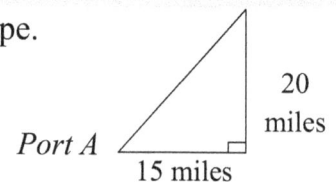

Port A, 20 miles, 15 miles

25) **Answer: D.**

The amount of money for x bookshelf is: 450x

Then, the total cost of all bookshelves is equal to: 450x + 840

The total cost, in dollar, per bookshelf is: $\frac{\text{Total cost}}{\text{number of items}} = \frac{450x+840}{x}$

Answers and Explanations

SSAT - Upper-Level

Practice Tests 1: Quantitative 2

1) Answer: E.

Let's review the options provided:

A. $24 \times \frac{1}{3} = \frac{24}{3} = 8 = 8$

B. $40 \times \frac{2}{10} = \frac{40}{5} = 8 = 8$

C. $5 \times \frac{16}{10} = \frac{80}{10} = 8 = 8$

D. $7 \times \frac{8}{7} = \frac{56}{7} = 8 = 8$

E. $12 \times \frac{5}{12} = \frac{5}{1} = 5 \neq 8$

2) Answer: C.

$\frac{53,238}{325} \cong 163.8092 \cong 164$

3) Answer: E.

Alex's mark is k less than Jason's mark. Then, from the choices provided Alex's mark can only be $23 - k$.

4) Answer: A.

The Area that one liter of paint is required: $52cm \times 100cm = 5,200cm^2$

Remember: $1\ m^2 = 10,000\ cm^2\ (100 \times 100 = 10,000)$,

Then $5,200cm^2 = 0.52\ m^2$

Number of liters of paint we need: $\frac{26}{0.52} = 50$ liters.

5) Answer: D.

$150 - 14\frac{19}{42} = (149 - 14) + \left(\frac{42}{42} - \frac{19}{42}\right) = 135\frac{23}{42}$

6) Answer: C.

Let x be the original price. If the price of the sofa is decreased by 30% to $455, then:

70% of $x = 455 \Rightarrow 0.70x = 455 \Rightarrow x = 455 \div 0.70 = 650$

7) Answer: E.

Number of times that the driver rests $= \dfrac{21}{7} = 3$

Driver's rest time $= 3$ hour and 10 minutes $= 190$ minutes

Then, 3×190 minutes $= 570$ minutes

1 hour $= 60$ minutes $\rightarrow 540$ minutes $= 9$ hours $\rightarrow 570$ minutes $= 9$ h and 30 min

8) Answer: E.

Find the difference of each pairs of numbers: $4, 5, 7, 10, 14, 19, 25, \_\_\_, 40$.

The difference of 4 and 5 is 1, 5 and 7 is 2, 7 and 10 is 3, 10 and 14 is 4, 14 and 19 is

5, 19 and 25 is 6, 25 and next number should be 7. The number is $25 + 7 = 32$

9) Answer: B.

Number of Mathematics book: $0.40 \times 650 = 260$

Number of English books: $0.20 \times 650 = 130$

Product of number of Mathematics and English books: $260 \times 130 = 33,800$

10) Answer: D.

The angle α is: $0.15 \times 360 = 54°$

The angle β is: $0.20 \times 360 = 72°$

11) Answer: D.

$x + 4 = 4 + 4 + 4 \rightarrow x = 8$

$y + 10 + 6 = 12 + 8 \rightarrow y + 16 = 20 \rightarrow y = 4$

Then, the perimeter is:

$4 + 12 + 4 + 4 + 8 + 10 + 4 + 6 + 4 + 8 = 64$

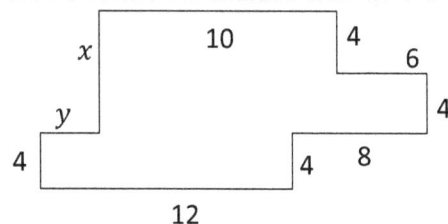

12) Answer: D.

$6y + 8 < 44 \rightarrow 6y < 44 - 8 \rightarrow 6y < 36 \rightarrow y < 6$

The only choice that is less than 6 is D.

13) Answer: D.

The capacity of a red box is 60% bigger than the capacity of a blue box and it can hold 77 books. Therefore, we want to find a number that 60% bigger than that number is 77. Let x be that number. Then: $1.60 \times x = 77$, Divide both sides of the equation by 1.6. Then: $x = \dfrac{77}{1.60} = 48.125 = 48$

14) Answer: C.

Since, E is the midpoint of AB, then the area of all triangles DAE, DEF, CFE and CBE are equal.

Let x be the area of one of the triangles, then: $4x = 200 \rightarrow x = 50$

The area of DEC $= 2x = 2(50) = 100$

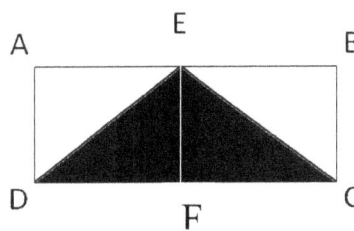

15) Answer: A.

$\frac{12+24}{2} = \frac{36}{2} = 18$ Then, $18 - 12 = 6$

16) Answer: B.

Amount of available petrol in tank: $51.8 - 8.4 - 27.6 + 16.54 = 32.34$ liters

17) Answer: C.

Let's review the choices provided:

A. $x = 2 \rightarrow$ The perimeter of the figure is: $4 + 6 + 4 + 2 + 2 = 18 \neq 24$

B. $x = 3 \rightarrow$ The perimeter of the figure is: $4 + 6 + 4 + 3 + 3 = 20 \neq 24$

C. $x = 5 \rightarrow$ The perimeter of the figure is: $4 + 6 + 4 + 5 + 5 = 24 = 24$

D. $x = 4 \rightarrow$ The perimeter of the figure is: $4 + 6 + 4 + 4 + 4 = 22 \neq 24$

E. $x = 1 \rightarrow$ The perimeter of the figure is: $4 + 6 + 4 + 1 + 1 = 16 \neq 24$

18) Answer: E.

We have two equations and three unknown variables, therefore x cannot be obtained.

19) Answer: E.

Let put some values for a and b. If $a = 9$ and $b = 2 \rightarrow a \times b = 18 \rightarrow \frac{18}{3} = 6 \rightarrow$ 18 is divisible by 3 then.

A. $5a + b = 45 + 2 = 47$ is not divisible by 3.

B. $2a + b = (2 \times 9) + 2 = 18 + 2 = 20$ is not divisible by 3.

C. $a + 6b = 9 + (6 \times 2) = 9 + 12 = 21$ is divisible by 3.

If $a = 17$ and $b = 3 \rightarrow a \times b = 51 \rightarrow \frac{51}{3} = 17$ is divisible by 3 then; $a + 6b = 17 + (6 \times 3) = 17 + 18 = 35$ is not divisible by 3.

D. $\frac{2a}{3b} = \frac{2 \times 17}{3 \times 3} = \frac{34}{9}$ is not divisible by 3

E. $6 \times 17 \times 3 = 306$

306 is divisible by 3. If you try any other numbers for a and b, you will get the same result.

20) Answer: D.

Perimeter of figure A is: $2\pi r = 2\pi \frac{10}{2} = 10\pi = 10 \times 3 = 30$

Area of figure B is: $5 \times 10 = 50$

Average $= \frac{30+50}{2} = \frac{80}{2} = 40$

21) Answer: B.

The angles on a straight line add up to 180 degrees. Let's review the choices provided: $(x + 30 + y + 2x + y = 3x + 2y + 30 = 180)$

A. $y = 15 \rightarrow 3x + 2y + 30 = 60 + 2(15) + 30 = 120 \neq 180$

B. $y = 45 \rightarrow 3x + 2y + 30 = 60 + 2(45) + 30 = 180$

C. $y = 30 \rightarrow 3x + 2y + 30 = 60 + 2(30) + 30 = 150 \neq 180$

D. $y = 10 \rightarrow 3x + 2y + 30 = 60 + 2(10) + 30 = 110 \neq 180$

E. $y = 40 \rightarrow 3x + 2y + 30 = 60 + 2(40) + 30 = 170 \neq 180$

22) Answer: E.

Let's review the choices provided:

A. $x = \frac{1}{6} \rightarrow \frac{3}{8} + \frac{1}{6} = \frac{9+4}{24} = \frac{13}{24} \cong 0.54 < 1$

B. $x = \frac{1}{5} \rightarrow \frac{3}{8} + \frac{1}{5} = \frac{15+8}{40} = \frac{23}{40} \cong 0.58 < 1$

C. $x = \frac{3}{7} \rightarrow \frac{3}{8} + \frac{3}{7} = \frac{21+24}{56} = \frac{45}{56} \cong 0.80 < 1$

D. $x = \frac{1}{9} \rightarrow \frac{3}{8} + \frac{1}{9} = \frac{27+8}{72} = \frac{35}{72} \cong 0.48 < 1$

E. $x = \frac{4}{5} \rightarrow \frac{3}{8} + \frac{4}{5} = \frac{15+32}{40} = \frac{47}{40} \cong 1.18 > 1$

Only choice E is correct.

23) Answer: C.

$565 \div 3 = \frac{565}{3} = \frac{500+60+5}{3} = \frac{500}{3} + \frac{60}{3} + \frac{5}{3}$

24) Answer: A.

$$\frac{5}{9} \times 45 = \frac{225}{9} = 25$$

25) Answer: C.

Set of numbers that are not composite between 1 and 10: A= {2, 3, 5, 7}

$$\text{Probability} \ = \ \frac{\text{number of desired outcomes}}{\text{number of total outcomes}} \ = \ \frac{4}{10} = \frac{2}{5}$$

Answers and Explanations

SSAT - Upper-Level

Practice Tests 2: Quantitative 1

1) **Answer: B.**

Digit 7 is in the tenths place.

2) **Answer: C.**

$0.06 \times 25.00 = \frac{6}{100} \times \frac{25}{1} = \frac{150}{100} = 1.5$

3) **Answer: C.**

Let x be the number. Write the equation and solve for x.

$\frac{2}{9} \times 45 = \frac{5}{7}x \rightarrow \frac{2 \times 45}{9} = \frac{5x}{7}$, use cross multiplication to solve for x.

$14 \times 45 = 9 \times 5x \Rightarrow 630 = 45x \Rightarrow x = 14$

4) **Answer: E.**

$x - 49 = -45 \rightarrow x = -45 + 49 \rightarrow x = 4$, Then; $x \times 3 = 4 \times 3 = 12$

5) **Answer: C.**

Number of packs needed equals to: $\frac{53}{6} \cong 8.83$

Then Mia must purchase 9 packs.

6) **Answer: D.**

His average speed was: $\frac{4.4}{0.2} = 22$ miles per hour.

7) **Answer: D.**

The amount of flour is: $x - 68$

8) **Answer: A.**

The time it takes to drive from city A to city B is: $\frac{6,758}{69} = 97.94$

It's approximately 98 hours.

9) **Answer: B.**

The perimeter of the quadrilateral is: $5 + 12 + 8 + 22 = 47$

10) **Answer: B.**

Number of males in the classroom is: $80 - 56 = 24$

Then, the percentage of males in the classroom is: $\frac{24}{80} \times 100 = 0.30 \times 100 = 30\%$

11) **Answer: D.**

Let x be the fifth quarter rate, then: $\frac{67+63+61+69+x}{5} = 68$

Multiply both sides of the above equation by 5. Then:

$5 \times \left(\frac{67+63+61+69+x}{5}\right) = 5 \times 68 \rightarrow 67 + 63 + 61 + 69 + x = 340 \rightarrow 260 + x = 340 \rightarrow x = 340 - 260 = 80$

12) **Answer: A.**

For one person the total cost is: $4.65

Therefore, for four persons, the total cost is: $4 \times \$4.65 = \$18.6$

13) **Answer: A.**

Use the information provided in the question to draw the shape.

Use Pythagorean Theorem: $a^2 + b^2 = c^2$

$15^2 + 36^2 = c^2 \Rightarrow 225 + 1{,}296 = c^2 \Rightarrow 1{,}521 = c^2$

$\Rightarrow c = 39$

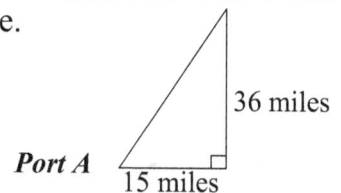

Port A   36 miles   15 miles

14) **Answer: E**

Percentage of men in city C $= \frac{780}{1{,}650} \times 100 = 47.27\%$

Percentage of women in city B $= \frac{410}{880} \times 100 = 46.59\%$

Percentage of men in city C to percentage of women in city B: $\frac{47.27}{46.59} = 1.015$

15) **Answer: D**

Ratio of men to women in city A: $\frac{650}{810} = 0.80$

Ratio of women to men in city B: $\frac{410}{470} = 0.87$

Ratio of women to men in city C: $\frac{780}{870} = 0.897$

Ratio of women to men in city D: $\frac{650}{700} = 0.93$

16) **Answer: C.**

Let the number of women should be added to city C be x, then:

$$\frac{780 + x}{870} = 1.5 \rightarrow 780 + x = 870 \times 1.5 = 1{,}305 \rightarrow x = 525$$

17) **Answer: C.**

Triple of 15,000 is 45,000. One ninth of them cancelled their tickets.

One ninth of 45,000 equal 5,000 ($\frac{1}{9} \times 45{,}000 = 5{,}000$).

40,000 (45,000 – 5,000 = 40,000) fans are attending this week

18) **Answer: A.**

The equation of a line in slope intercept form is: $y = mx + b$

Solve for y.

$$7x + y = 32 \Rightarrow y = -7x + 32$$

$$y = -7x + 32$$

The slope of this line is -7.

The slope of the line perpendicular to this line is:

$$m_1 \times m_2 = -1 \Rightarrow -7 \times m_2 = -1 \Rightarrow m_2 = \frac{1}{7}$$

19) **Answer: B.**

The length of MN is equal to: $5x + 7x = 12x$

Then: $12x = 24 \rightarrow x = \frac{24}{12} = 2$

The length of ON is equal to: $7x = 7 \times 2 = 14$ cm.

20) **Answer: A.**

$$\frac{x}{3+8} = \frac{y}{22-14} \rightarrow \frac{x}{11} = \frac{y}{8} \rightarrow 8x = 11y \rightarrow y = \frac{8}{11}x$$

21) **Answer: A.**

$$\sqrt[6]{x^{55}} = \sqrt[6]{x^{54} \times x} = \sqrt[6]{x^{54}} \times \sqrt[6]{x} = x^{\frac{54}{6}} \times \sqrt[6]{x} = x^9 \sqrt[6]{x}$$

22) **Answer: C.**

The average speed of John is: $240 \div 8 = 30$ km

The average speed of Alice is: $420 \div 6 = 70$ km

Write the ratio and simplify. $30: 70 \Rightarrow 3: 7$

23) **Answer: D.**

Smallest 4–digit number is 1,000, and biggest 4–digit number is 9,999. The difference is: 8,999.

24) **Answer: D.**

Perimeter of rectangle A is equal to: $2 \times (8 + 4) = 2 \times 12 = 24$

Perimeter of rectangle B is equal to: $2 \times (3 + 7) = 2 \times 10 = 20$

Therefore: $\frac{20}{24} \times 100 = 0.83 \times 100 = 83\%$

25) **Answer: A.**

Let x be the number of years. Therefore, $5,000 per year equals 5,000x.

Starting from $28,000 annual salary means you should add that amount to 5,000x.

Income more than that is: $I > 5,000 x + 28,000$

Answers and Explanations

SSAT - Upper-Level

Practice Tests 2: Quantitative 2

1) **Answer: A.**

$$0.46 \times 11.7 = \frac{46}{100} \times \frac{117}{10} = \frac{46 \times 117}{100 \times 10} = \frac{5,382}{1,000} = 5.382$$

2) **Answer: C.**

$$3\frac{1}{5} \times 4\frac{1}{2} = \frac{16}{5} \times \frac{9}{2} = \frac{16 \times 9}{5 \times 2} = \frac{144}{10} = 14\frac{2}{5}$$

3) **Answer: C.**

Let x be the number, then; $\frac{2}{5}x = 52 \rightarrow x = \frac{5 \times 52}{2} = 130$

Therefore: $\frac{7}{10}x = \frac{7}{10} \times 130 = 91$

4) **Answer: E.**

A. $\frac{5}{6} \times \frac{2}{7} = \frac{5}{21}$ is not equal to whole number.

B. $\frac{1}{4} + \frac{1}{5} = \frac{5+4}{20} = \frac{9}{20}$ is not equal to whole number.

C. $\frac{115}{15} = \frac{23}{3} = 7.66$ is not equal to whole number.

D. $5.4 + 3 = 8.4$ is not equal to whole number.

E. $4.5 + \frac{7}{2} = 4.5 + 3.5 = 8$ is a whole number.

5) **Answer: E.**

Number of boxes equal to: $\frac{360}{18} = \frac{840}{2} = 20$

6) **Answer: D.**

The question is this: 360.10 is what percent of 554?

Use percent formula: Part $= \frac{\text{percent}}{100} \times$ whole.

$360.10 = \frac{\text{percent}}{100} \times 554 \rightarrow 360.10 = \frac{\text{percent} \times 554}{100} \rightarrow 36,010 = \text{percent} \times 554,$

Then, Percent $= \frac{36,010}{554} = 65$

360.10 is 65 % of 554. Therefore, the discount is: $100\% - 65\% = 35\%$

7) Answer: C.

Use distance formula:

Distance = Rate × time ⇒ 500 = 40 × T, divide both sides by 50. $\frac{500}{40} = T \Rightarrow T = 12.5$ hours.

Change hours to minutes for the decimal part. 0.5 hours = 0.5 × 60 = 30 minutes

8) Answer: A.

Use formula of rectangle prism volume.

V = (length) (width) (height) ⇒ 5,600 = (14) (5) (height) ⇒ height = 5,600 ÷ 70 = 80

9) Answer: D.

Use simple interest formula:

I = prt (I = interest, p = principal, r = rate, t = time)

I = (27,000)(0.0245)(6) = $3,969

10) Answer: D.

7 cubed is: 7 × 7 × 7 = 49 × 7 = 343

11) Answer: E.

The angle x and 58 are complementary angles. Therefore:

$x + 58 = 180 \Rightarrow x = 180° - 58° \Rightarrow x = 122°$

12) Answer: C.

The rate of construction company $= \frac{20 \text{ cm}}{1 \text{ min}} = 20$ cm/min

Height of the wall after 30 min $= \frac{20 \text{ cm}}{1 \text{ min}} \times 30$ min = 600 cm

Let x be the height of wall, then $\frac{4}{5}x = 600$ cm $\rightarrow x = \frac{5 \times 600}{4}$

$\rightarrow x = 7,50$ cm = 7.5 m

13) Answer: E.

$y = 5ab + 3b^3$

Plug in the values of a and b in the equation: a = 4 and b = 3

$y = 5 (4) (3) + 3 (3)^3 = 60 + 3(27) = 60 + 81 = 141$

14) Answer: C.

Let's review the choices provided:

A. $56 - 8 = 48 \to \frac{48}{8} = 6 \neq 7$

B. $56 - 11 = 45 \to \frac{45}{11} \neq 7$

C. $56 - 7 = 49 \to \frac{49}{7} = 7 = 7$

D. $56 - 22 = 34 \to \frac{34}{22} = \frac{17}{11} \neq 7$

E. $56 - 5 = 51 \to \frac{51}{5} \neq 7$

15) Answer: E.

Speed of car A is: $\frac{350}{14} = 25$ Km/h

Speed of car B is: $\frac{350}{7} = 50$ Km/h $\to 50 - 25 = 25$ Km/h

16) Answer: E.

The ratio of boys to girls is 7:2. Therefore, there are 7 boys out of 9 students. To find the answer, first divide the total number of students by 9, then multiply the result by 7.

$540 \div 9 = 60 \Rightarrow 60 \times 7 = 420$

17) Answer: A.

$\frac{(9+5)^2}{7} + 6 = \frac{(14)^2}{7} + 6 = \frac{196}{7} + 6 = 28 + 6 = 34$

18) Answer: C.

Use the formula of the area of circles.

Area $= \pi r^2 \Rightarrow 25\pi = \pi r^2 \Rightarrow 25 = r^2 \Rightarrow r = 5$

Radius of the circle is 5. Now, use the circumference formula:

Circumference $= 2\pi r = 2\pi (5) = 10\pi$

19) Answer: D.

$0.6x = (0.4) \times 15 \to x = 10 \to (x + 5)^2 = (10 + 5)^2 = (15)^2 = 225$

20) Answer: C.

Prime factorizing of $24 = 2 \times 2 \times 2 \times 3$

Prime factorizing of $72 = 2 \times 2 \times 2 \times 3 \times 3$

GCF$= 2 \times 2 \times 2 \times 3 = 24$

21) Answer: C.

Length of the rectangle is: $\frac{5}{9} \times 36 = 20$

Perimeter of rectangle is: $2 \times (36 + 20) = 112$

22) Answer: B.

$\frac{4}{7}$ Of $147 = \frac{4}{7} \times 147 = 84$

$\frac{1}{4}$ Of $84 = \frac{1}{4} \times 84 = 21$

23) Answer: B.

The sum of all angles in a quadrilateral is 360 degrees.

Let x be the smallest angle in the quadrilateral. Then the angles are: $3x, 4x, 5x, 6x$

$3x + 4x + 5x + 6x = 360 \rightarrow 18x = 360 \rightarrow x = 20$

The angles in the quadrilateral are $60°, 80°, 100°,$ and $120°$

The smallest angle is 60 degrees.

24) Answer: C.

x is the number of all sales profit and 5% of it is:

$5\% \times x = 0.05x$

Employee's revenue: $0.05x + 8,000$

25) Answer: A.

$\frac{3}{11} = 0.272; \quad \frac{4}{7} \cong 0.571; \quad \frac{9}{13} \cong 0.69; \quad \frac{4}{5} = 0.80$

"End"